IMAGES
of America

SAN FRANCISCO'S MIDWINTER EXPOSITION

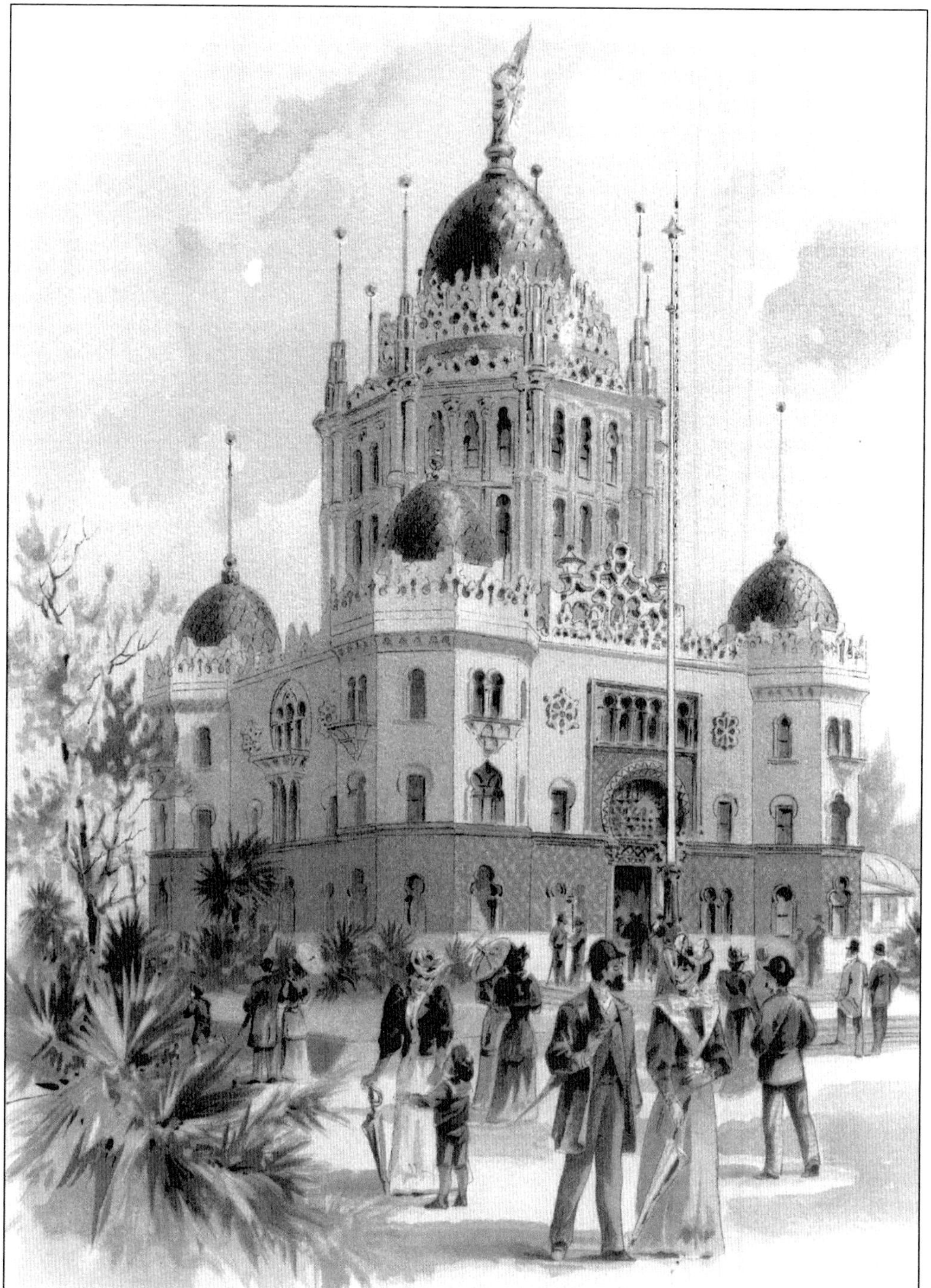

The Administration Building of the Midwinter Fair, an outstanding example of Victorian wedding cake architecture, set the tone for the entire exposition: eclectic, exciting, eccentric, and marvelously whimsical. No visitors seeing it could doubt that wonders awaited them at the first international exhibition held west of the Mississippi.

IMAGES
of America

SAN FRANCISCO'S MIDWINTER EXPOSITION

Dr. William Lipsky

ARCADIA
PUBLISHING

ISBN 978-0-7385-2088-9

Published by Arcadia Publishing
Charleston, South Carolina

Printed in the United States of America

Library of Congress Catalog Card Number: Applied For.

For all general information contact Arcadia Publishing at:
Telephone 843-853-2070
Fax 843-853-0044
E-Mail sales@arcadiapublishing.com
For customer service and orders:
Toll-Free 1-888-313-2665

Visit us on the Internet at www.arcadiapublishing.com

This book is for Don Price.

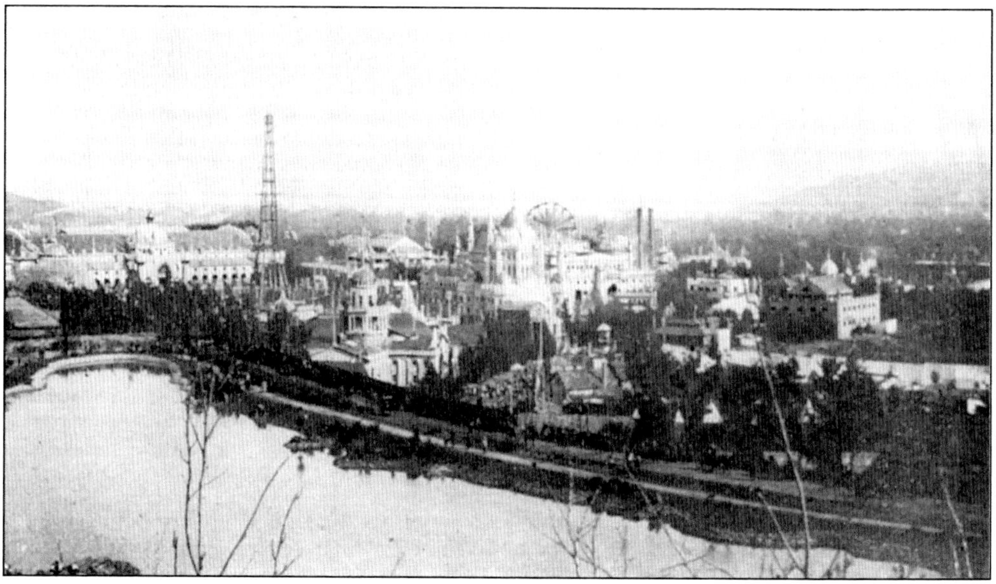

The fairgrounds from the observatory atop Strawberry Hill showed a compactness remarkable for world's fairs. On the left, the Manufactures Building, with its huge arched glass roof, is clearly visible. Barely to the right of it stands the Electric Tower, the tallest structure at the exposition. The Administration Building is dead center, with the Firth Wheel behind it a little to the right.

Contents

Acknowledgments

Many thanks to Ken Prag for making his extensive collection of San Francisco and Midwinter Fair images available without hesitation or restriction.

Photographs on pages 13a, 20a, 58a and b, 67a, 78b, 83b, 84a, and 104b are in the collection of the San Francisco History Center, San Francisco Public Library, and used with kind permission.

A Note on the Illustrations

The Midwinter Fair came and went just as Thomas Edison was completing his first successful experiments with motion pictures. Only still images of the exposition exist, the greater number taken by Isaiah Taber, the Fair's official photographer. Many of the images here were taken by him. Amateur photographers were allowed to take pictures at the Fair, but only for their personal use. Images 57b, 81b, 105b, 107b, and 128b were taken by hobbyists and have never before been published. A few images are from stereocards published by B.W. Kilburn.

We are indebted to Taber and to all the photographers of the Midwinter Fair, professional and amateur, for preserving it in pictures through which we may look into the past from time to time.

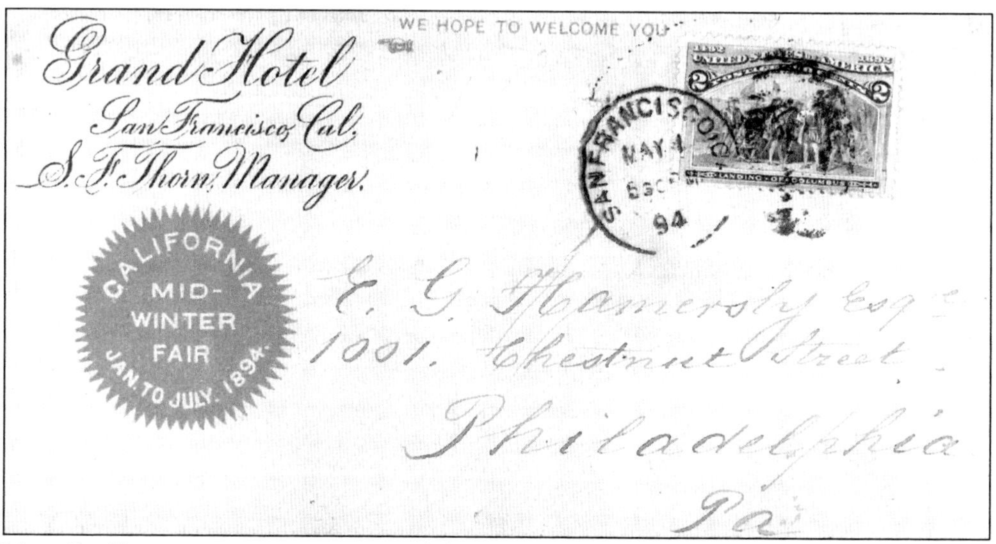

Many San Francisco companies promoted the Midwinter Fair by adding a little something about it to their business correspondence, usually a stock cut like this one on an envelope from the Grand Hotel.

Introduction

In the winter of 1894, while the rest of the country shivered in the frigid weather, San Francisco gave a party and invited the world. Why face snow and sleet in the East? Why suffer blizzards in the Midwest or tempests in the Rockies? Come to California, "The Land of Sunshine, Fruit, and Flowers." Come to the Fair. Kith and kind took notice and more than two million people answered the call. The California Midwinter International Exposition, the first American world's fair held west of the Mississippi River, was a rousing success.

The idea for a midwinter fair occurred not in San Francisco, but in Chicago. When the World's Columbian Exposition opened on May 1, 1893, Michael de Young, publisher of the San Francisco *Chronicle*, saw an opportunity for his city in Chicago's stellar accomplishment. When it closed in October, why not bring the best of the Columbian's exhibits and attractions to California for a Western world's fair?

Like the rest of the country, San Francisco was experiencing a period of economic downturn just then, the beginnings of the 19th century's worst depression. Here was a way to improve conditions. The enterprise would create jobs and stimulate the economy. It would allow local residents who couldn't afford the three-and-a-half day train trip to Chicago to enjoy some of the attractions. It would bring visitors to San Francisco and it would showcase the City as a winter destination, "where people of the frozen East could find a genial climate without abandoning metropolitan pleasures." At the same time, it would promote California's agriculture, horticulture, viticulture, forestry, mining, and manufacture to the world.

De Young saw other good reasons to have a world's fair in San Francisco. With the exhibits already in the United States, already paid for, it would be inexpensive to produce. If it used unimproved land in Golden Gate Park, which it could have rent free, it also would be inexpensive to operate. Roads and trolley lines already were in place. Expertise was readily available in Chicago. Promotion was easily obtained through his newspaper. If it concentrated a large number of visitors into a short period of time—say no more than 60 days—the fair might even make a profit.

On May 31, still in Chicago, de Young met with visiting California business leaders to propose this commercial world's fair for the coming winter. Some newspapers of the day would call it "a typical Californian idea," but de Young's core group liked it. After all, the Columbian exposition had cost something like $28 million to produce, not including private construction, exhibits, attractions, stands, salaries, and goods. They could launch a Western fair for little more than the cost of shipping the displays from Chicago—and they wouldn't have to pay the bills of lading. The gathering agreed to go forward with the project and sent telegrams to the mayor of San Francisco and the governor of California to see what they thought of it.

Mayor Levi Ellert liked the idea, too. On June 5, at de Young's request, he assembled key local merchants to ask for their support. Most were indifferent. It required a second Chicago meeting, on June 11, to move the project forward, specifically de Young's announcement that he had raised $41,500 as earnest money for the exposition and, as important, had obtained the commitments of important foreign commissioners to bring their exhibits to California.

That did it. On June 13, another, larger meeting in San Francisco endorsed the project; the next day the mayor appointed a Citizens Committee of 50 to create a business plan. Presented and passed on June 29, their plan made one major change to the original proposal: the organizers decided the fair should be general in nature, not merely commercial. On July 10, the Finance Committee announced its simple strategy to raise the capital for the exposition: it would ask the public for donations. Thirty days later it had enough money pledged to set the date for groundbreaking, which took place on August 24. Five months later, on January 27, 1894, the Fair formally opened.

The California Midwinter International Exposition had gone from fantasy to fact in less than eight months, financed without any federal, state, or local government bonds, loans, grants, or subsidies of any kind. The city's business leaders put up the seed money, but that was hardly enough. Additional funds came through appeals to companies, organizations, and clubs. Many individuals contributed a day's wages or a percentage of their salaries. The various campaigns raised $344,319.59, including $2.74 donated by the city's newsboys. There was gossip that some workers were coerced into being civic minded, but, if true, most San Franciscans gave willingly; with no income tax, nobody needed a tax deduction. Contributors didn't expect to receive any money back and the Fair received no requests for repayment when it closed.

Typical of the times, all department heads—and all their assistants—were men. The Chicago Exposition had employed female managers—who, after all, would dare deny Mrs. Potter Palmer anything?—but only of female enterprises; the Midwinter Fair, having no Women's Building, Women's Department, or Women's Section, had no thought of women in responsible positions.

Not everything was in place the first day, but given that the Columbian Exposition, although larger, had taken four years to create, the achievement was nothing short of astounding; not everything had been completed by opening day at the Chicago fair, either. While the East and Midwest suffered through yet another cold season, San Francisco, enjoying its warm winter days, invited the rest of the country—and the world—to enjoy them, too. All they need do was come to the Fair.

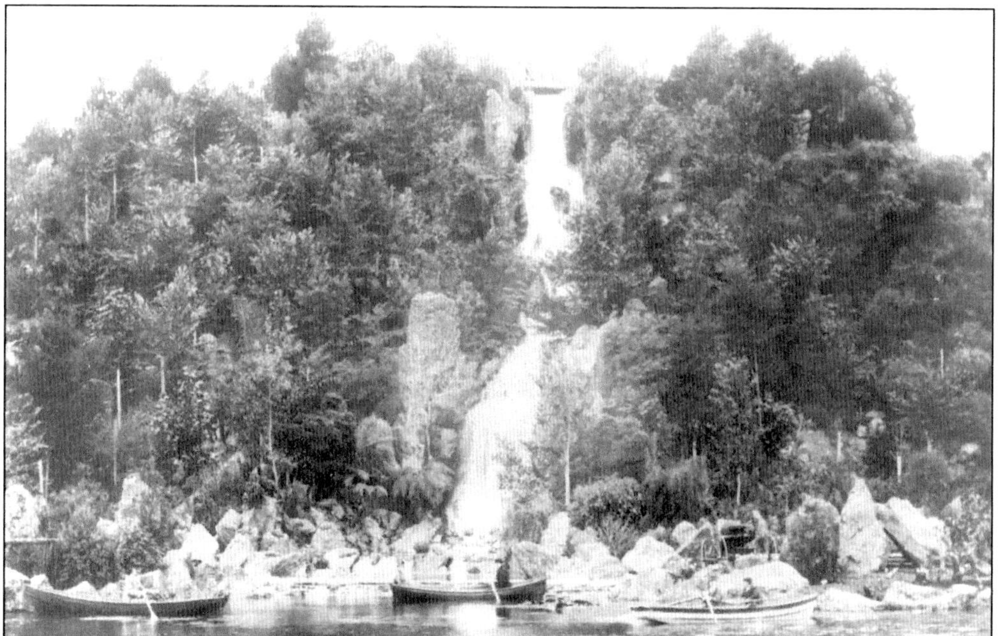

Boating on Golden Gate Park's Stow Lake was a favorite pastime of the 1890s, when this photograph was taken. Huntington Falls, center, with a drop of 125 feet, was dedicated on April 9, 1894. Elm, maple, beech, linden, and other shade trees graced the hillside, sheltering more than 10,000 ferns. Both were created from the sand dunes of the park.

One

A Joyous Celebration

The idea of a midwinter fair in San Francisco proved irresistible. Michael de Young initially had envisioned a local commercial fair, not an international exposition. He thought it might operate for 30 or possibly 60 days and use no more than 60 acres of park land. Instead it stayed open 6 months and spread over almost 200 acres. It had exhibits from every county in California, almost every state, and 37 foreign countries. Even poets and composers got caught up in excitement:

Have you lately been over to Golden Gate Park
Where there's something to please from daylight to dark?
But the thing that most pleases just now I declare
Is the site of the wonderful Midwinter Fair.

Where late was a wilderness lonely and drear
A vision of palaces now doth appear
With spires and with towers resplendent and rare
Adorning the site of the Midwinter Fair.

They are coming from Germany, Holland, and Spain
From Austria, France, and the Lombardy Plain
With laces and velvets and fabrics of hair
The ladies to please at the Midwinter Fair.

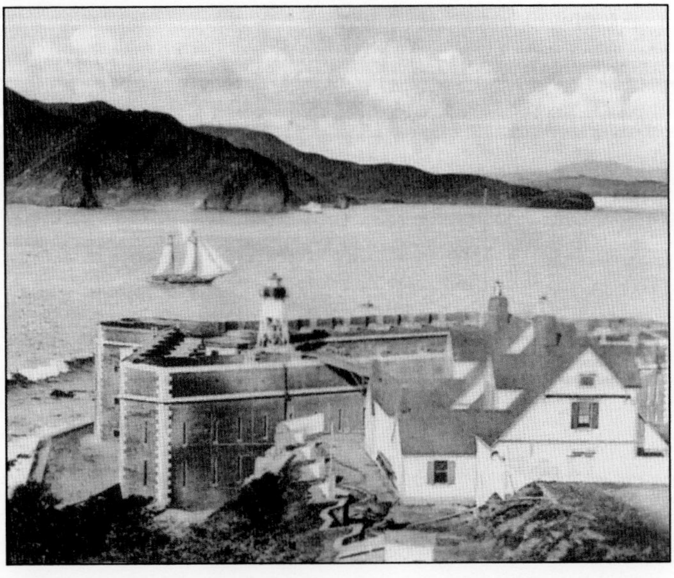

Visitors to San Francisco in 1894 who came by sea, then as now, passed through the Golden Gate. Fort Point, seen in the foreground, had been constructed by the U.S. Army Corps of Engineers between 1853 and 1861 to prevent hostile natives from entering the bay. Alcatraz, designated the official U.S. Army prison for the western states and territories in 1861, wouldn't become a Federal penitentiary for another 40 years. The Golden Gate Bridge was 43 years in the future.

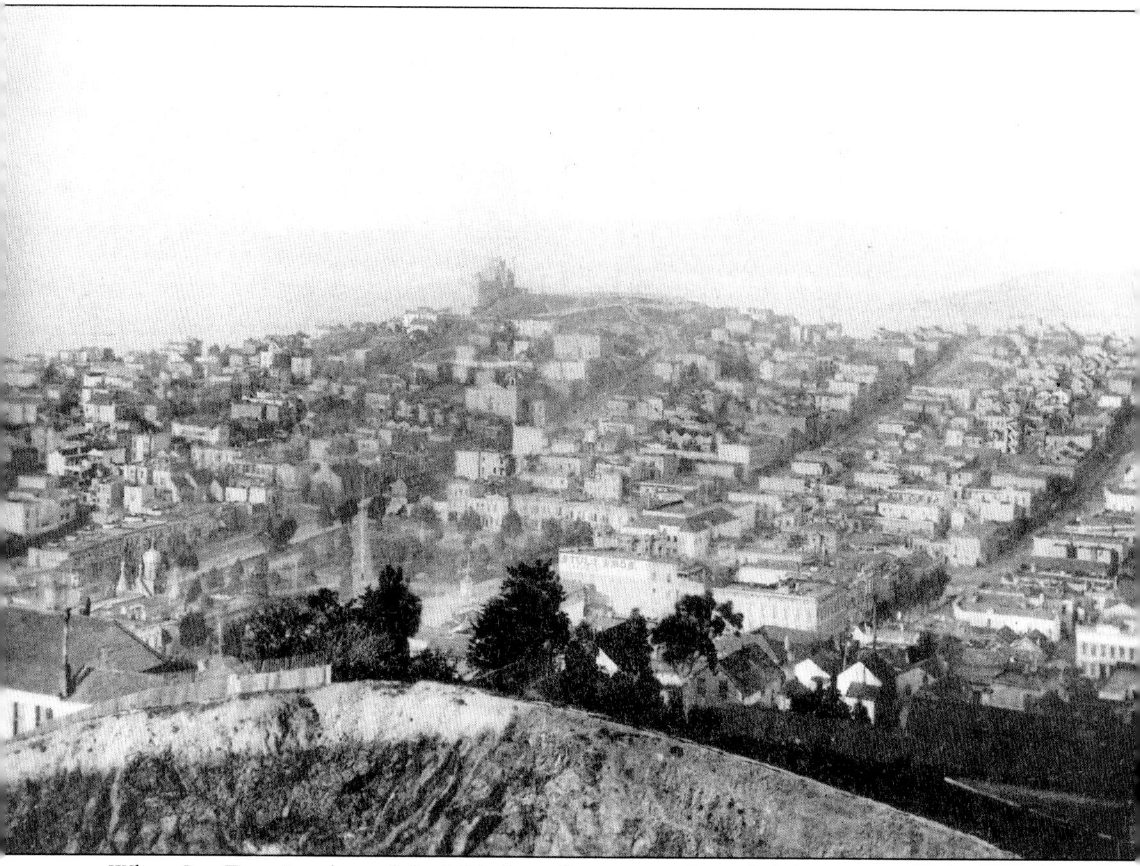

When San Francisco hosted the Midwinter Fair, it was the eighth largest city in the U.S., home to more than 300,000 people. It no longer was the city that Mark Twain, Ada Issacs Menken, and Bret Harte had known: rugged, open, bursting with energy. It wasn't even the city of Norton I, self-proclaimed Emperor of the U.S. and Protector of Mexico, who died in 1880 after a reign of some 20 years. It was now the financial, cultural, and social capital of the West, where dignified gentlemen and genteel ladies could be trusted to handle a silver dining service properly. The city, seen here from Russian Hill, boasted of a widely diverse population, but in fact more than 90 percent of the citizenry was white. Chinese residents, the largest minority, made up less than 9 percent of the community; African-Americans, less than 1 percent; Japanese, 0.2 percent.

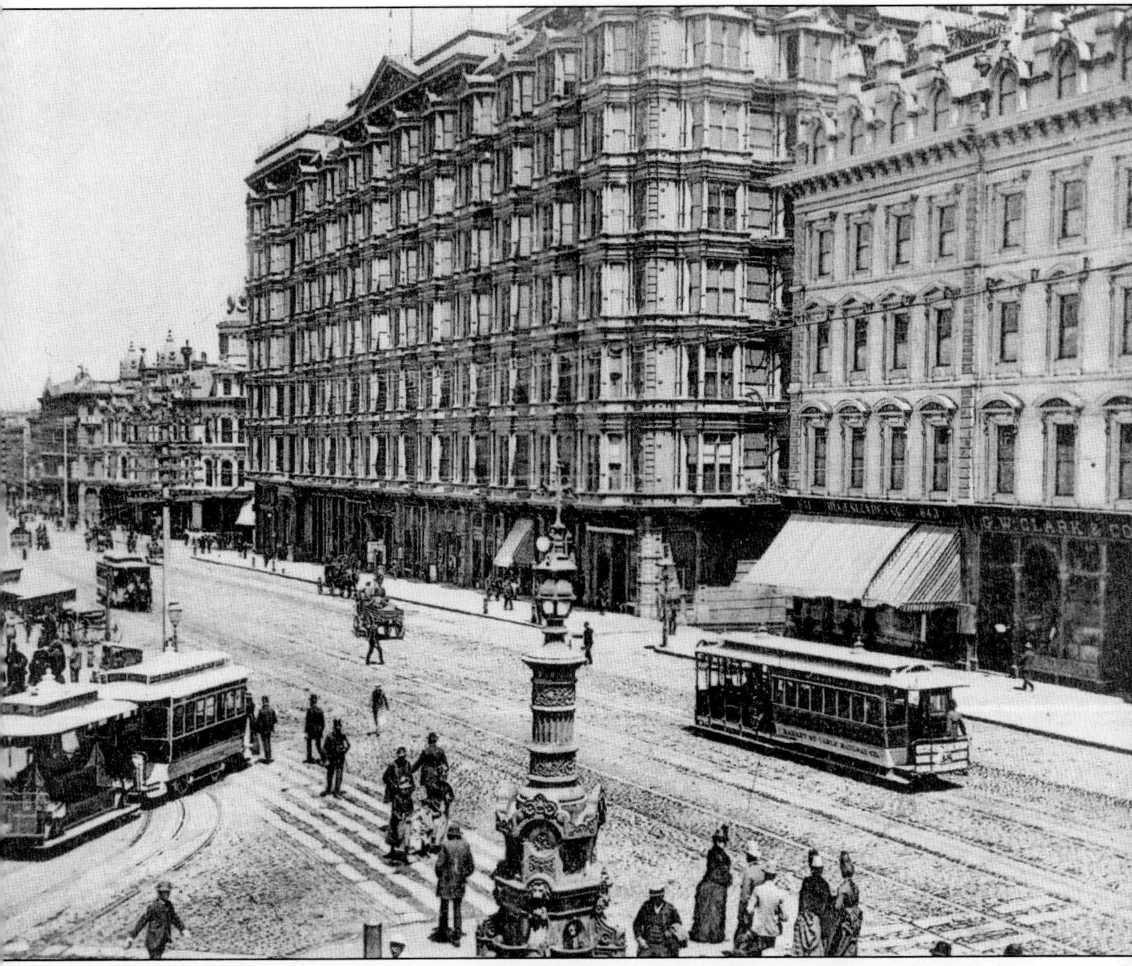

In 1894, as in 1849, people from all over the world came to San Francisco. It was the city that adored Lotta Crabtree, whose gift of a fountain, shown here, still stands on Market Street; the city that drew Elfro Jefferson, who became California's first African-American millionaire; that read local author Ambrose Bierce's column, "The Prattler," in the San Francisco *Examiner*; that mourned the passing of Robert Louis Stevenson, who died the year of the Fair; enjoyed performances by Bert Williams and George Walker, who first worked together at Jack Halahan's Cramorne Theater on Market Street, later named the Midway Plaisance; and cherished Gelett Burgess, who, the next year, published his immortal quatrain in *The Lark*, a local literary journal he co-edited:

I never saw a Purple Cow,
I never hope to see one.
But I can tell you anyhow,
I'd rather see than be one.

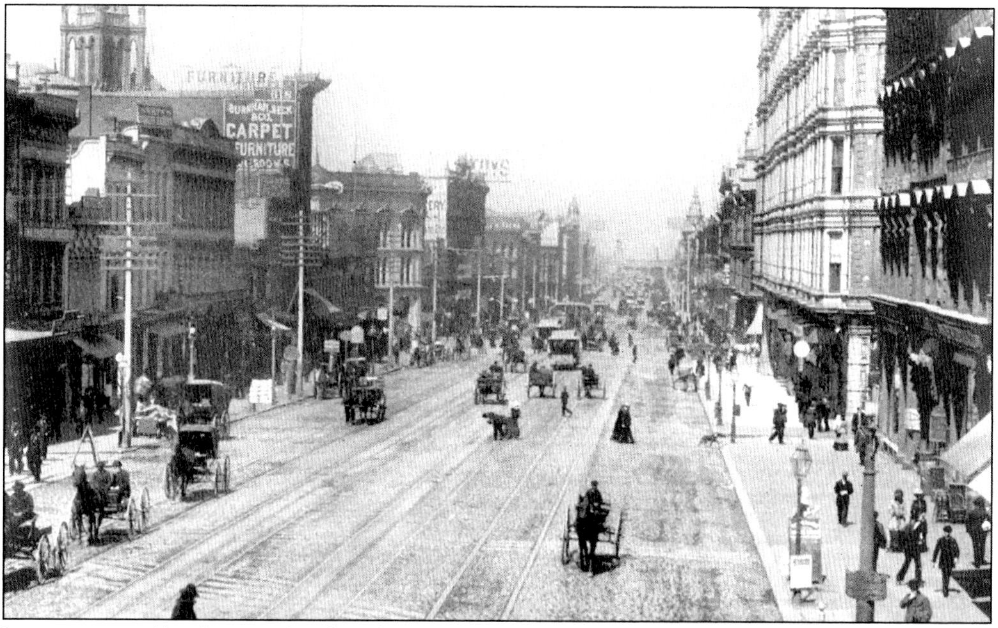

It was a city of banking and railroading, entrepreneurs and nabobs, working men and working women. In its theaters, patrons heard Zélie de Lussan, a great Carmen of her day; minstrel star (and soon to be recording pioneer) Richard J. Jose; and Luisa Tetrazzini, a particular favorite. In its bookstalls, readers found Rudyard Kipling's *The Jungle Book* and Mark Twain's *Puddn'head Wilson*, both published the year of the Fair.

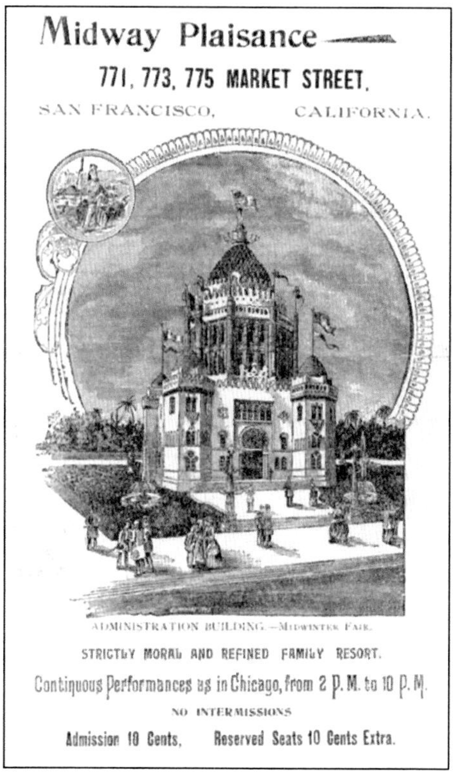

San Francisco had popular amusements, too. It also had a saloon for every 96 citizens. The Midway Plaisance Theatre, one of the city's cheapest variety houses, still claimed to be a "strictly moral and refined family resort" where audiences could see celebrated glass eaters, female wood sawing contests, "Eulelia the Artistic Oriental Dancer" and Lucretia, "the most wonderful artist extant."

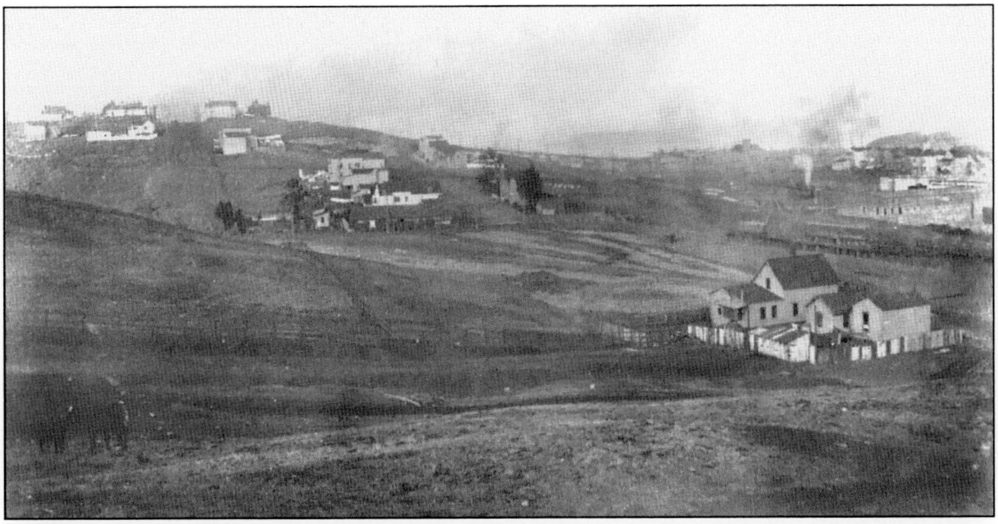

Suburban communities such as Sunnyside, Ocean View, and the Portrero, shown here, were gaining population rapidly. Of course, some weren't happy with the changes. "The Potrero is not what is used to be," said one resident. "A few years ago, I knew every person in the Potrero. Now the Potrero has put on city airs and I don't know whether I am acquainted with my next door neighbor or not."

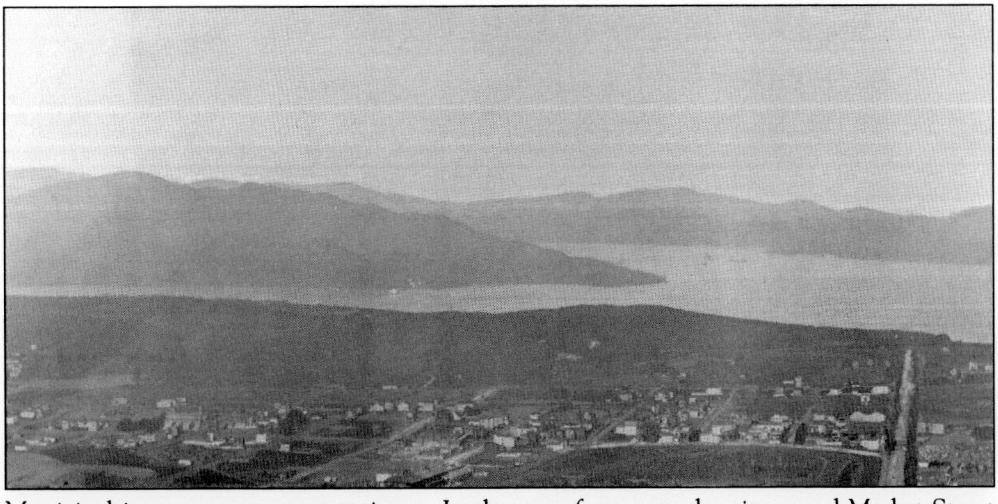

Municipal improvements were eminent. In the next few years, the city paved Market Street from the junction of Valencia and Haight Streets all the way to Seventeenth and Castro Streets and macadamized J Street south of Golden Gate Park, among others. The area north of the park, shown here, was connected to the rest of the city by two cable railways: the McAllister Street line and the Geary Street line.

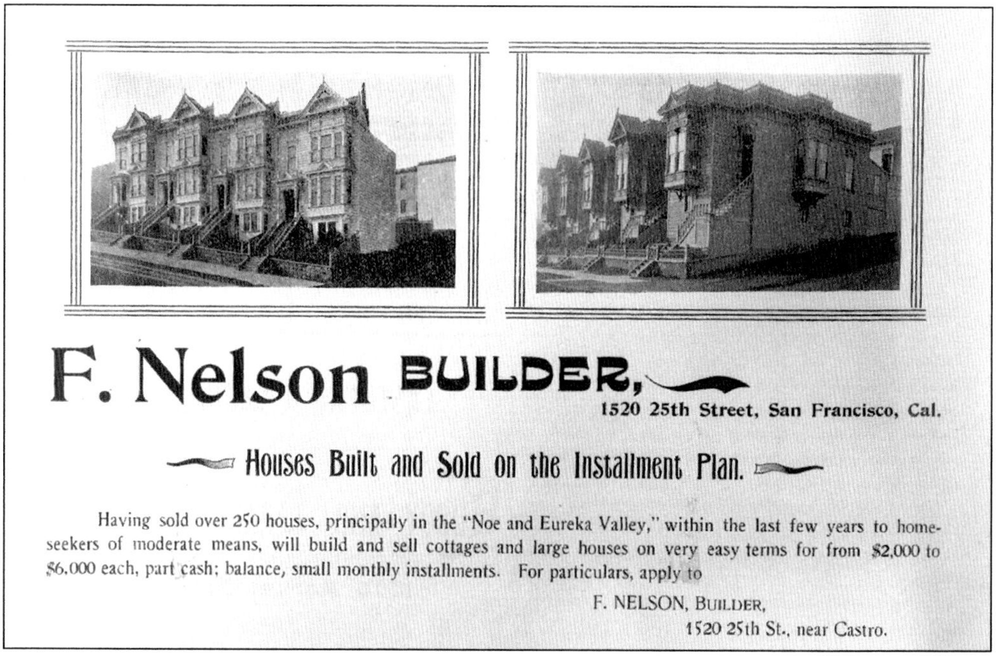

The City was expensive even then. People could buy a new townhouse in Eureka or Noe Valley for $2,000–$6,000—less than a monthly mortgage payment for a home in the same neighborhoods today—but the average annual household income in 1894 was about $400. A family could survive on that if it shopped carefully, although, to make ends meet, some heated their homes with driftwood they found along the shore.

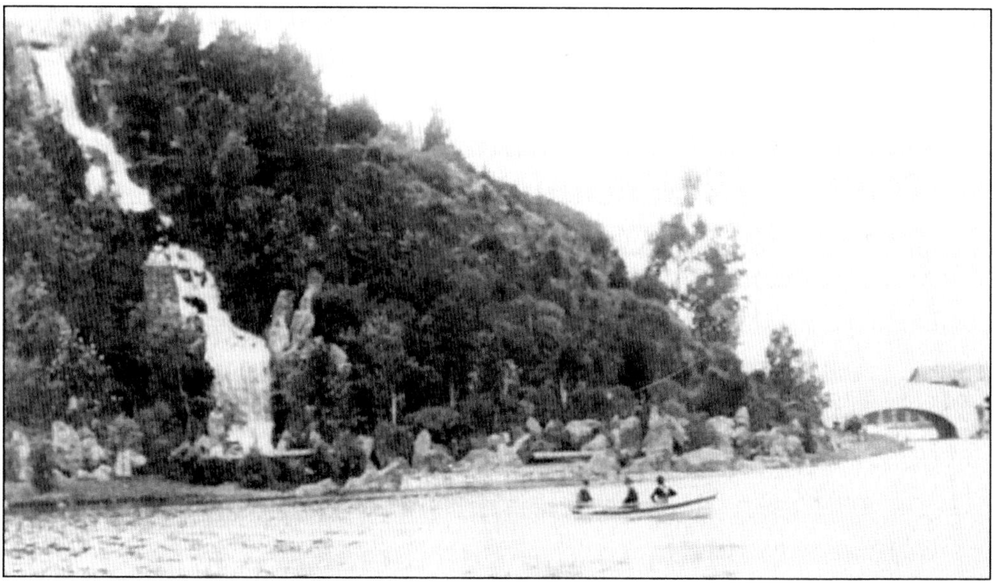

Not everyone was happy with the idea of a Midwinter Fair. Some called it the "Midwinter Fake" when promised jobs went only to union members. The city's park commissioners didn't oppose the fair itself, only the proposal to put it in Golden Gate Park. Neither they nor Park Superintendent John McLaren were please to have their newly planted greenery, such as here at Stow Lake, trampled by several million visitors.

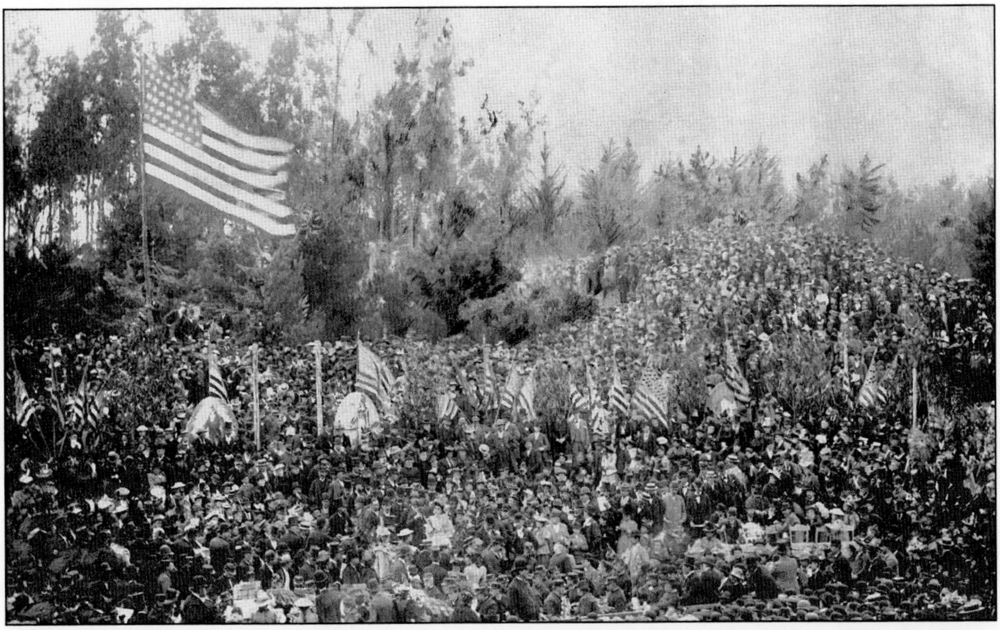

Groundbreaking ceremonies took place on August 24. Sixty thousand people gathered to listen to speeches and a potpourri of national airs. De Young, now Director-General of the Exposition, said a few words, then turned over a spadeful of soil. The action began five months of hectic work to transform a 200-acre patch of sand dunes and scrub brush into the Sunset City by January 1, 1894, opening day.

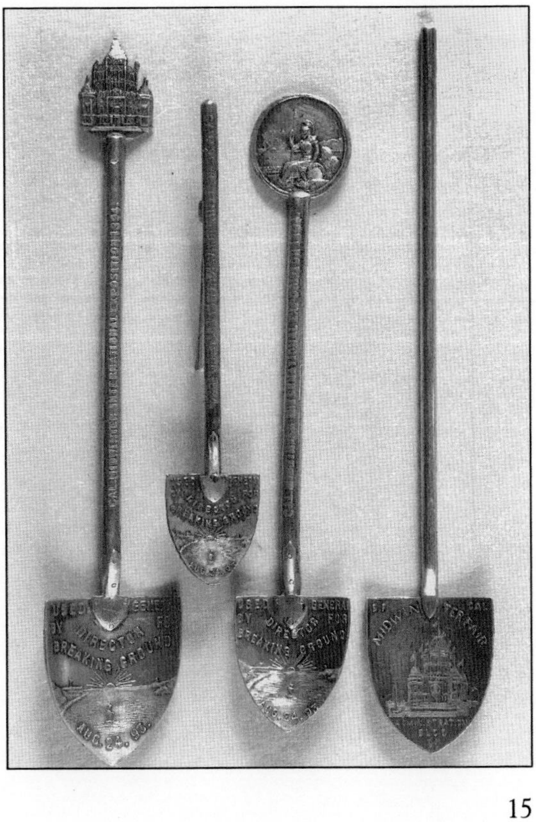

The silver shovel used for groundbreaking soon appeared in miniature as souvenir spoons and pins. The original was auctioned during the ceremony to raise money for construction; the Golden Rule Bazaar paid $105 for it, a true bargain. The first spadeful of earth went to men's clothier Adolph Roos for $650, but it had been salted with a handful of rare coins.

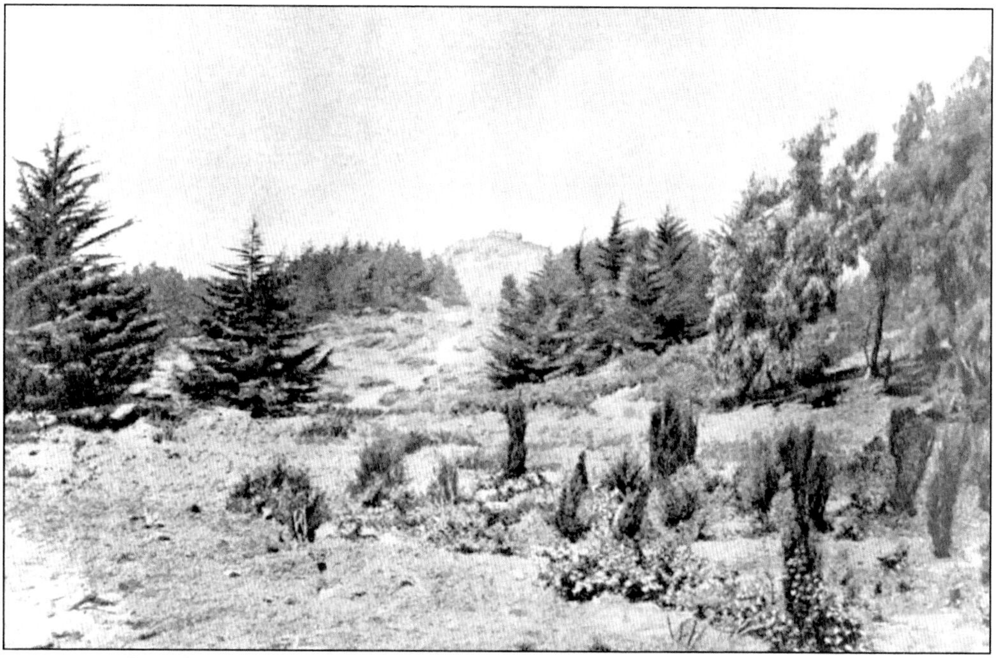

Concert Valley, the site chosen for the Fair, was an area of Golden Gate Park that had yet to be landscaped. More than any other event, the Fair crystallized the city's view of what the park would be: a place of civilized nature in an urban landscape, designed to refresh the mind as well as the soul; it wouldn't be a natural environment.

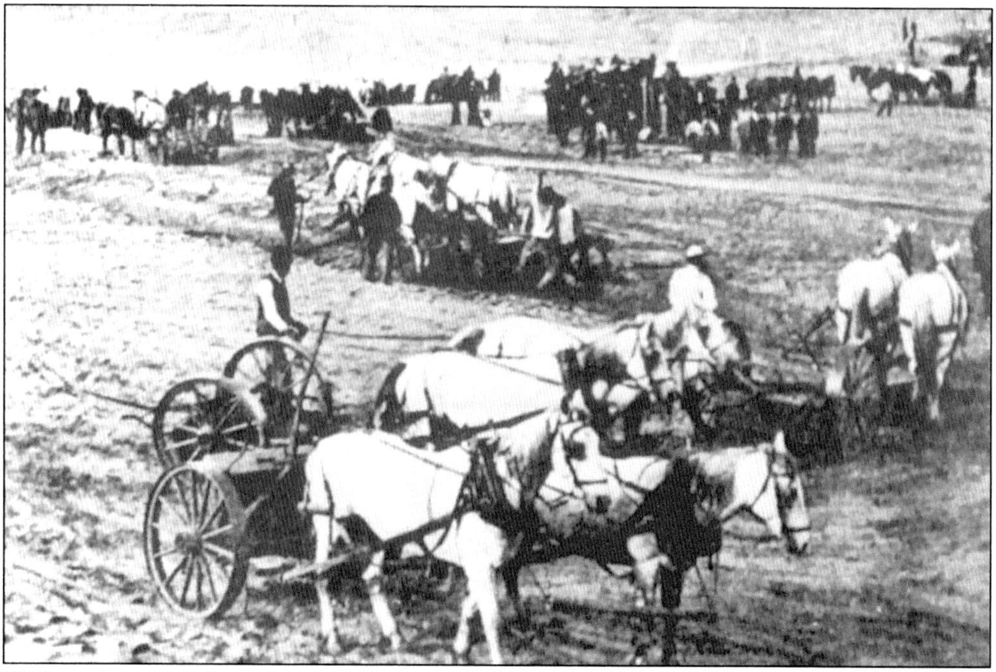

Without heavy earth-moving machines, workers, using plows and scrapers pulled by teams of horses, began preparing the site even as the groundbreaking ceremony proceeded. They soon were followed by carpenters, craftsmen, landscapists, colorists, painters, and sculptors.

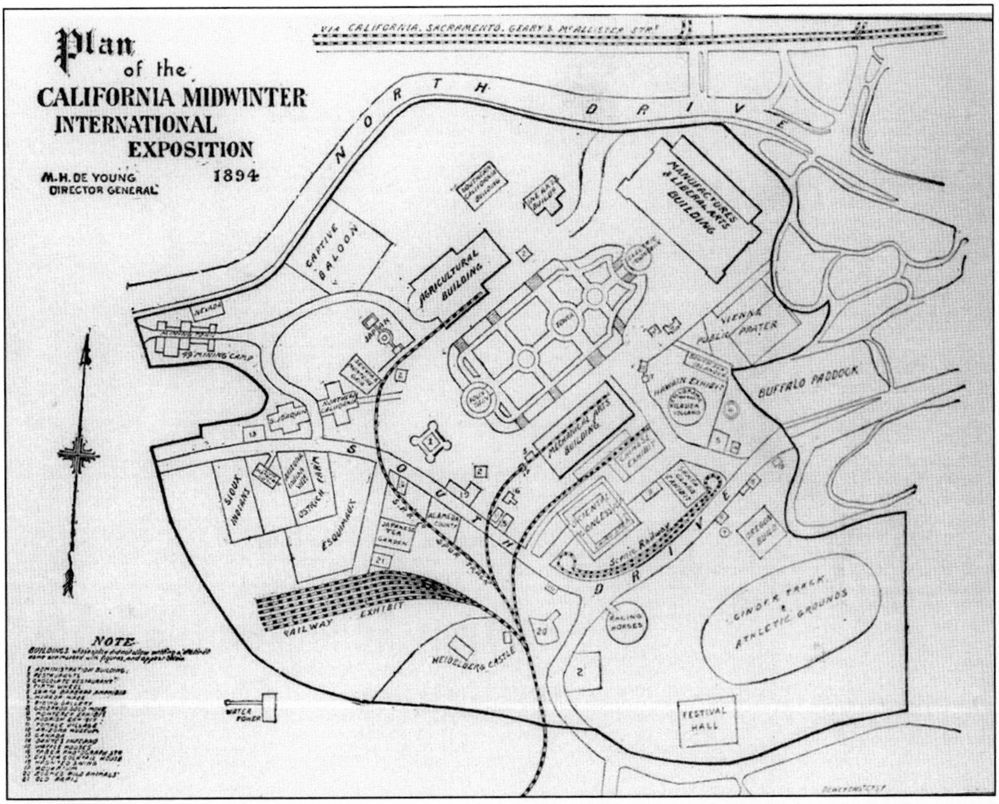

The fairgrounds were designed as two concentric circles. The main exhibit halls were built around the inner circle; other attractions were scattered about the outer. The task of actually building the exposition went to chief engineer Michael O'Shaughnessy. He laid out the grounds, sited the buildings, and devised the infrastructure. Later, as city engineer of San Francisco, he built the system that still brings the city its water from the Sierra-Nevada Mountains.

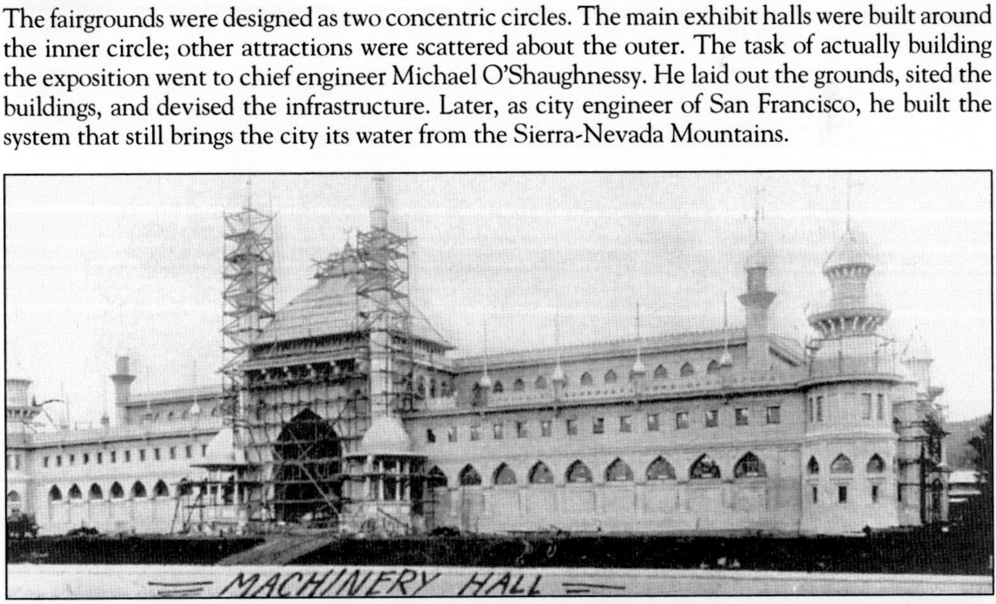

Construction began the moment grading had been completed. Workers earned about $2 a day, a substantial amount when the average paycheck was about $7.50 a week. In other parts of the country, the '90s were a decade of violent labor-management confrontations, but the Fair hired only union members. Because the unions admitted white men exclusively, no African-Americans, Asians, or women got work building the Sunset City.

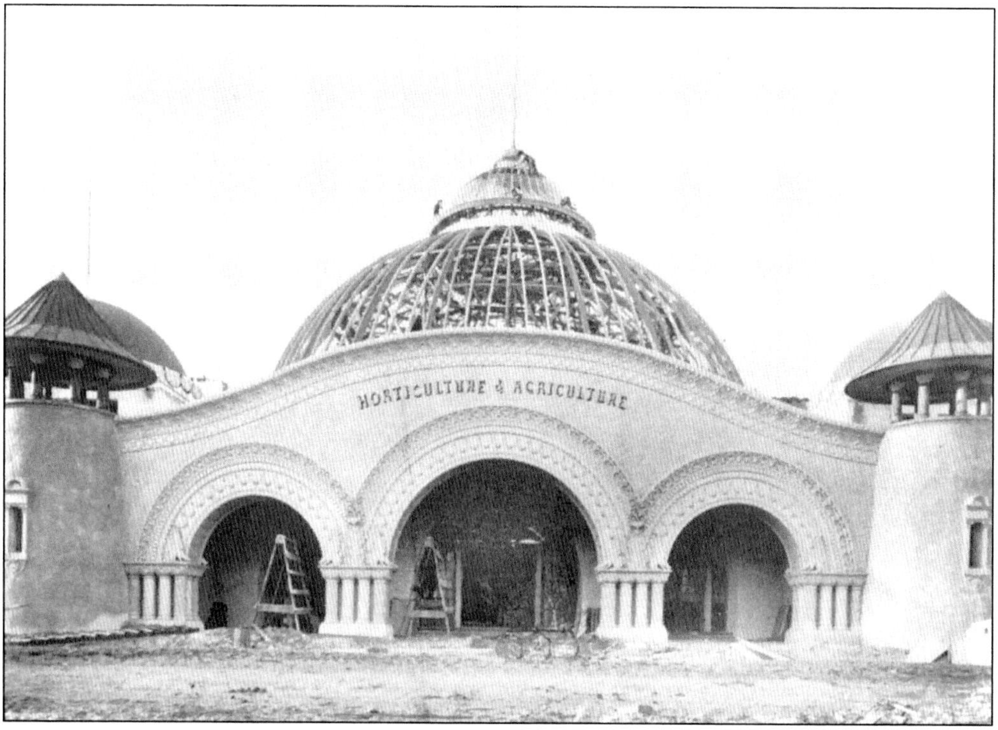

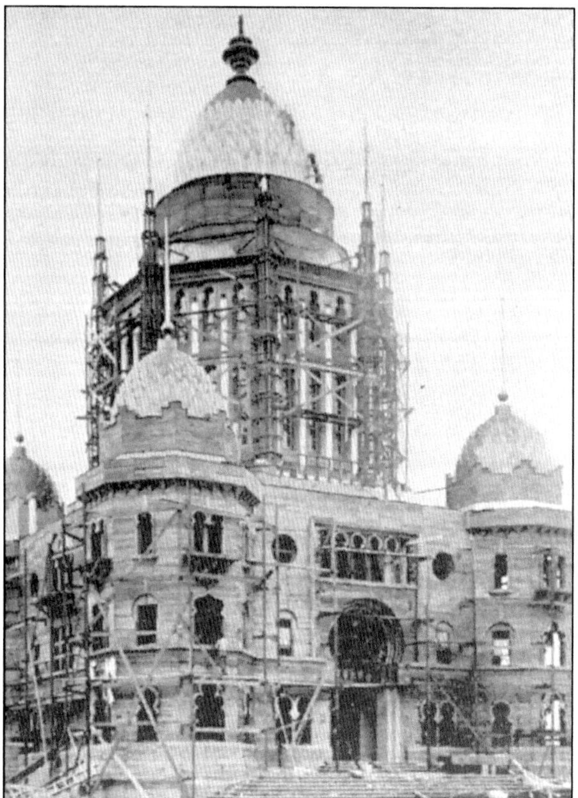

Under construction, the Horticulture and Agriculture Building looked like it was being built to last a hundred years. It wasn't. Only the Fine Arts Building was to be permanent. The others were fabricated from wallboard attached to wood and iron frames, then covered with staff, a mix of plaster, fiber, and cement. Kate Greenaway's whimsical decorations on either side of the main archway are yet to be added.

On December 15, 1893, with only two weeks remaining before opening day, the Administration Building appeared as if it would never be finished in time, which was so. The grand opening was postponed to January 27. By the middle of December, however, with enough work done, the exposition company allowed visitors, at 25¢ each, to tour the grounds. They watched hundreds of men planting trees, carting lumber, and installing benches.

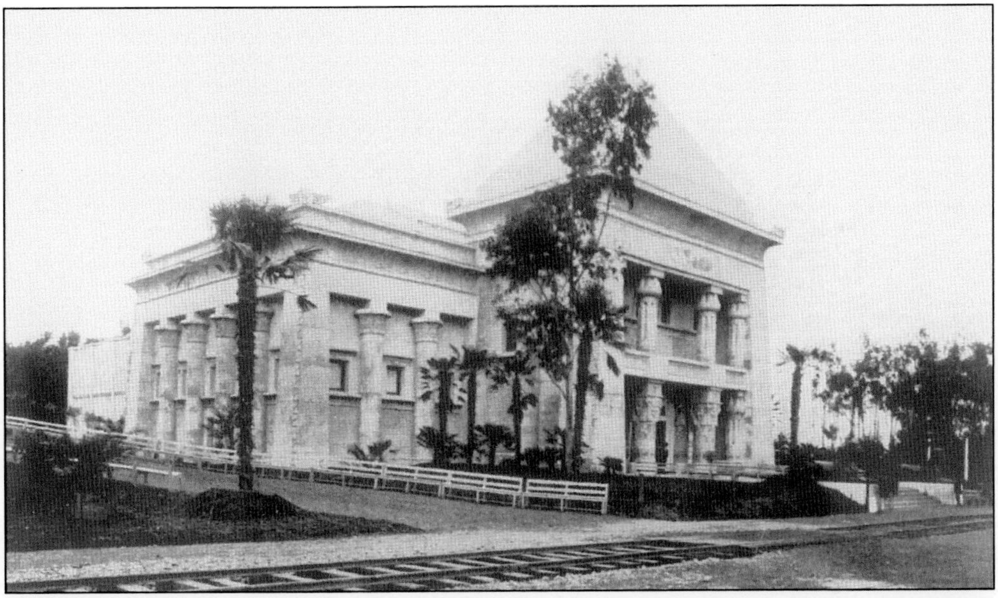

The Fine Arts Building, which became San Francisco's first municipal art gallery, was completed right on schedule, although here its surroundings still need some attention. The original design called for sculpted elephants' heads at the bases of the second story pillars, but they didn't happen; disappointed fairgoers made do with the elephant heads on the Chocolate Pavilion. By opening day, workers had planted additional greenery and buried the railway track.

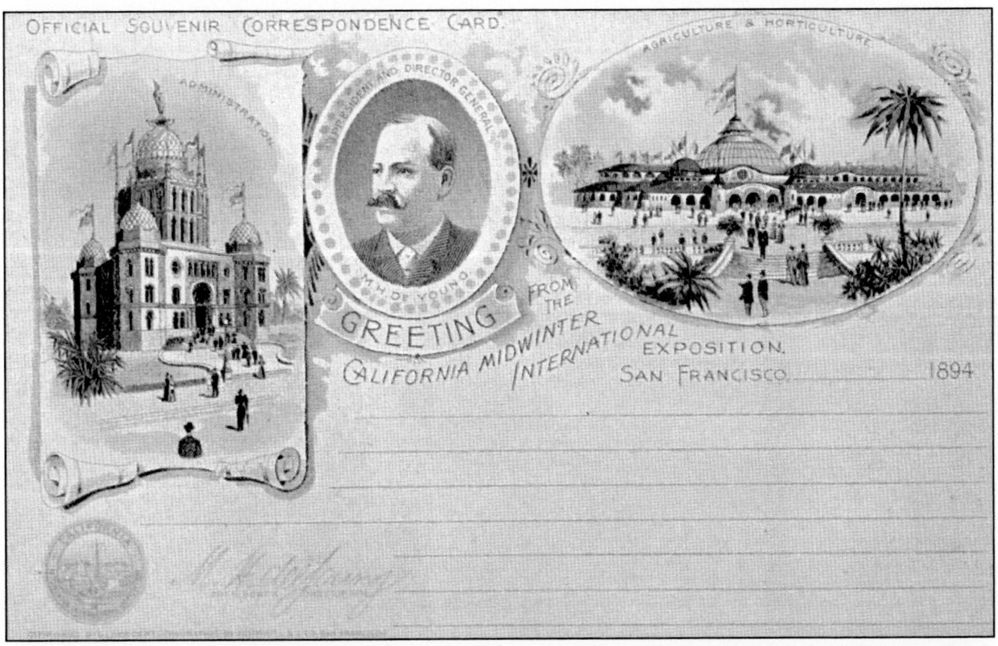

The exposition sanctioned two sets of souvenir post cards, and entrepreneurs produced a third, unauthorized set. The U.S. Government issued no commemorative stamps or coins for the Midwinter Fair, even with a branch mint in San Francisco. It made up for the oversight in 1915, when it created more face and denomination different coins for the Panama-Pacific Exposition than for any other world's fair.

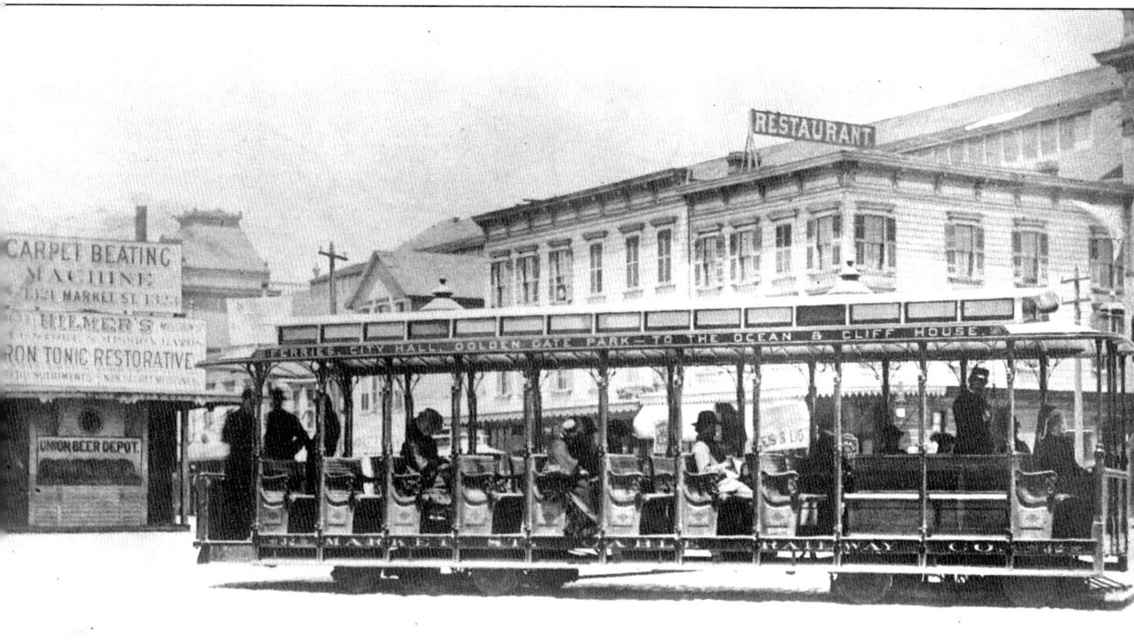

Fairgoers used streetcars, cable cars, or horse cars to reach the pay gates. The cost from anywhere in the city was 5¢. They also could hire a cab or a hack, or use their own horse-drawn vehicle, if they had one. Blizzards and snows might dog the East, but San Franciscans rode to Golden Gate Park in open streetcars or cable cars like this one at Market and Larkin Streets, operated by the Market Street Railway Company Street. Cooled on a warm day by nature's own air conditioning and offering an unrestricted view of the passing celebration of life, it remains the finest form of urban transportation ever invented. A new type of transportation was coming, though. The year after the Fair, Charles Duryea received the first U.S. patent given to an American for a gasoline-powered automobile. Soon after a Duryea Motor Wagon collided with a bicycle rider, causing the first automobile accident in the U.S.

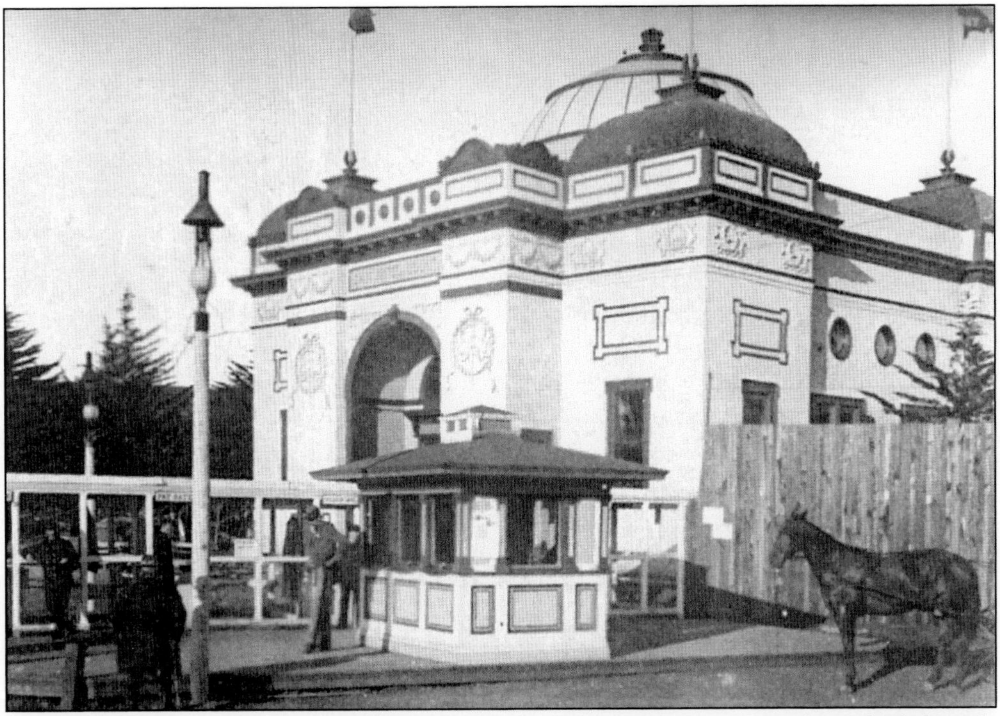

Visitors entered the Fair at one of three locations. The South Gate, seen here, had the sole bicycle entrance. All ticket booths, turnstiles, and exit doors came from the Columbian Exposition. The grounds were open from 7 a.m. until 11 p.m.; buildings were accessible from 8 a.m. until 10 p.m. A board fence 12 feet high and 11,350 feet long, topped with four strands of barbed wire, surrounded it all.

Dedication Day, set for January 1, was postponed, ironically, due to bad weather in the rest of the country. First, exhibits couldn't be shipped because Eastern blizzards prevented rolling stock from reaching Chicago. Then they couldn't be delivered because Sierra snowstorms halted rail traffic into California. It became a blessing in disguise: it was still unfinished. Informal ceremonies were held on January 1, with the formal opening rescheduled for January 27.

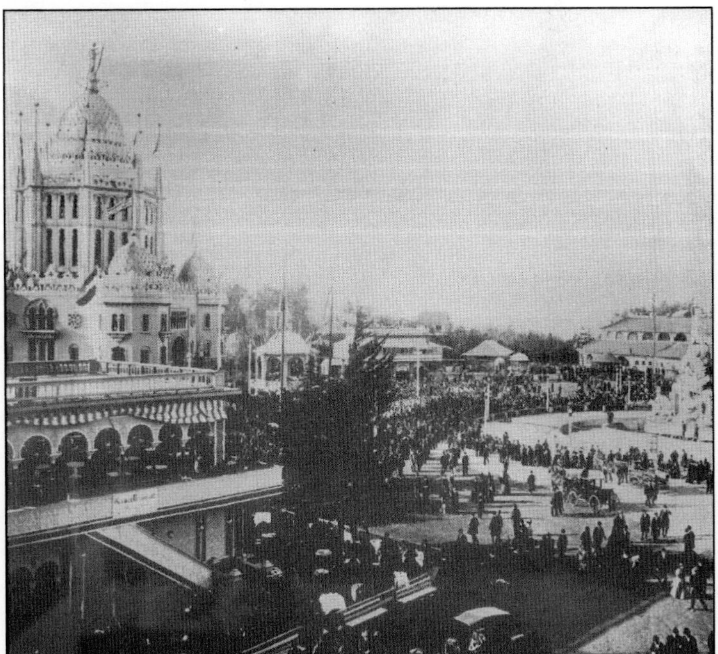

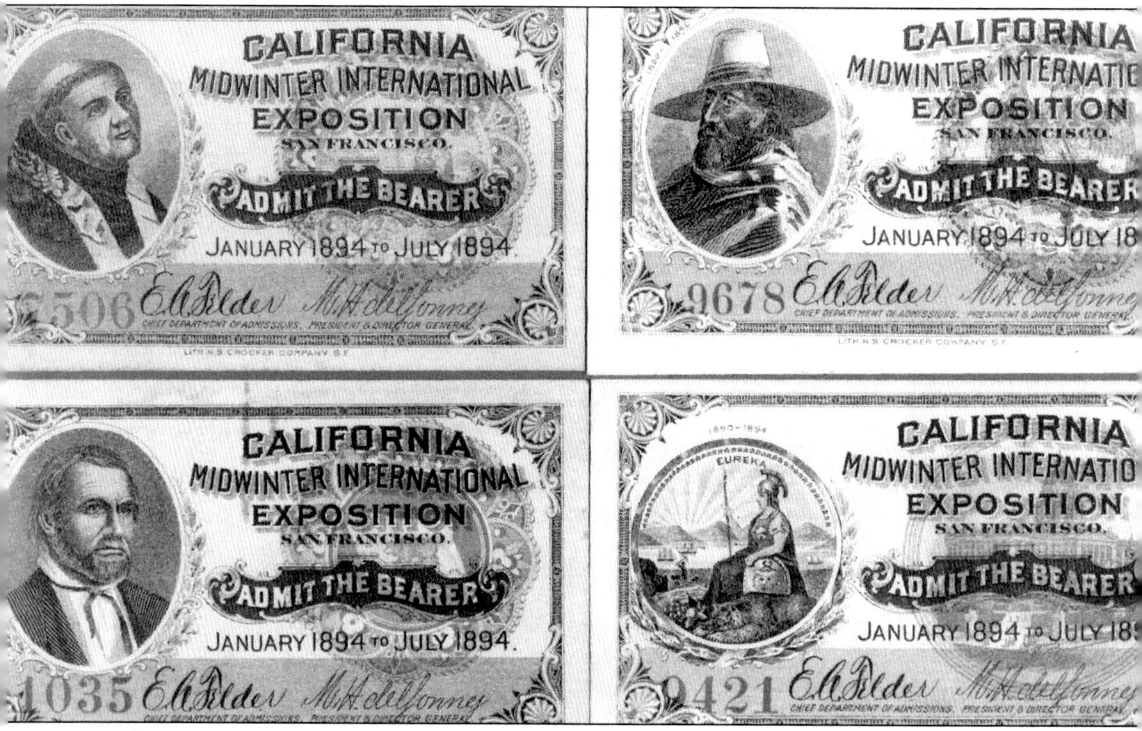

The Fair issued four souvenir admission tickets, representing the four flags of California history. Franciscan missionary Junipero Serra personified the era of Spanish colonialism (1768–1822). A cabillero embodied the period of Mexico's rule (1822–1846). A pioneer characterized the time of the Gold Rush and the Bear Republic. The Great Seal of California symbolized the period of statehood beginning in 1850. Admission was 50¢ for adults, 25¢ for children, with both reduced in early May to 25¢ and 10¢. The Fair also sold discount ticket books of 100 admissions for $40, but stopped when sidewalk brokers began reselling individual tickets from them for more than the discount, but less than the full price. Organizations, congresses, conventions, and other bulk ticket buyers received day passes.

As owner of the San Francisco *Chronicle*, de Young certainly recognized the importance of press coverage of the Fair. So too did many of the concessionaires. The Exposition's press pass admitted a member of the fourth estate to the fairgrounds, but not to any of the paid concessions. That required a second pass like this one for the '49 Camp Theater and dance hall.

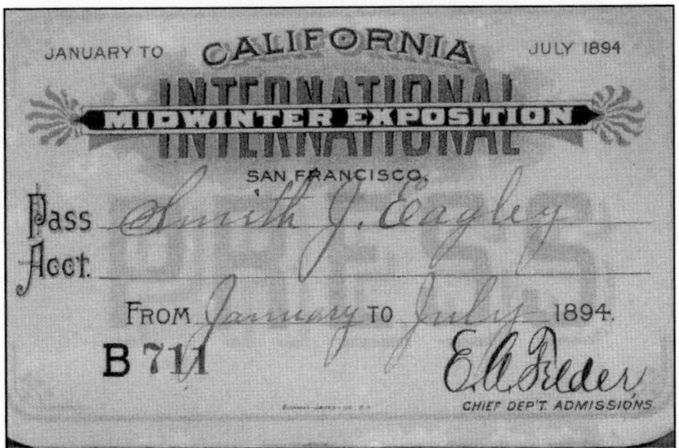

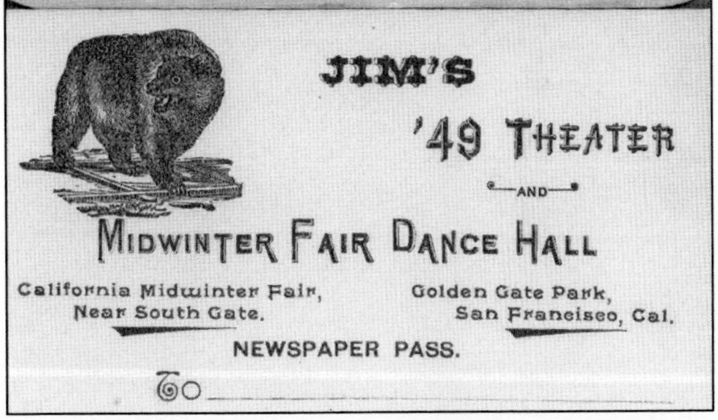

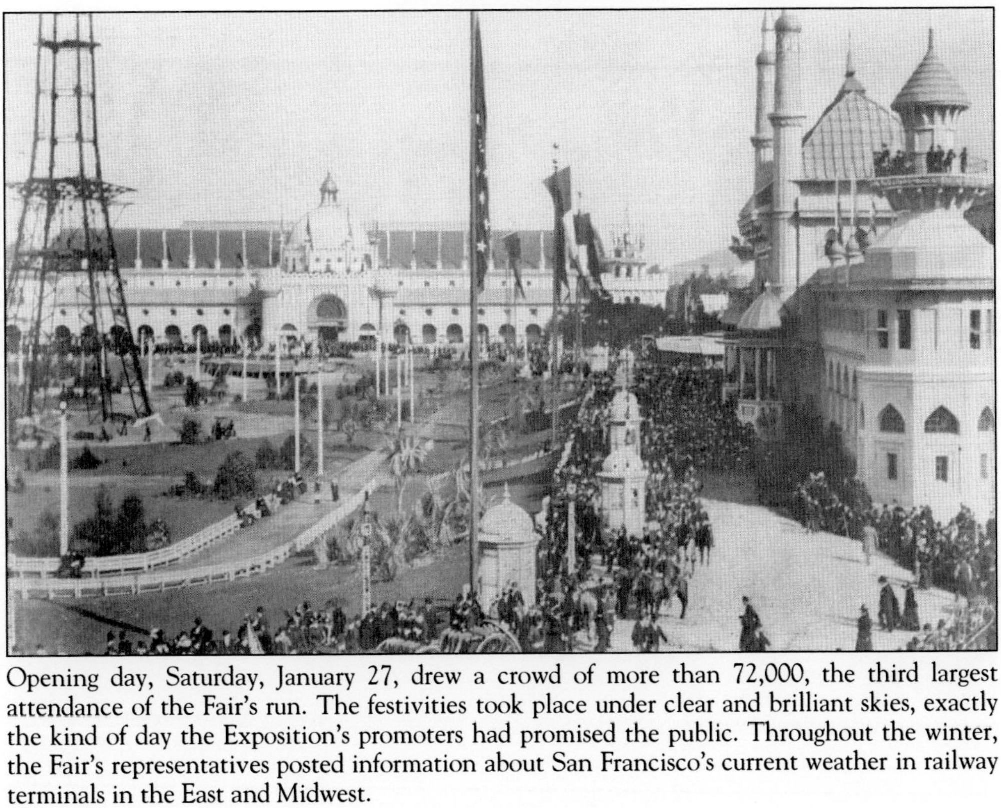

Opening day, Saturday, January 27, drew a crowd of more than 72,000, the third largest attendance of the Fair's run. The festivities took place under clear and brilliant skies, exactly the kind of day the Exposition's promoters had promised the public. Throughout the winter, the Fair's representatives posted information about San Francisco's current weather in railway terminals in the East and Midwest.

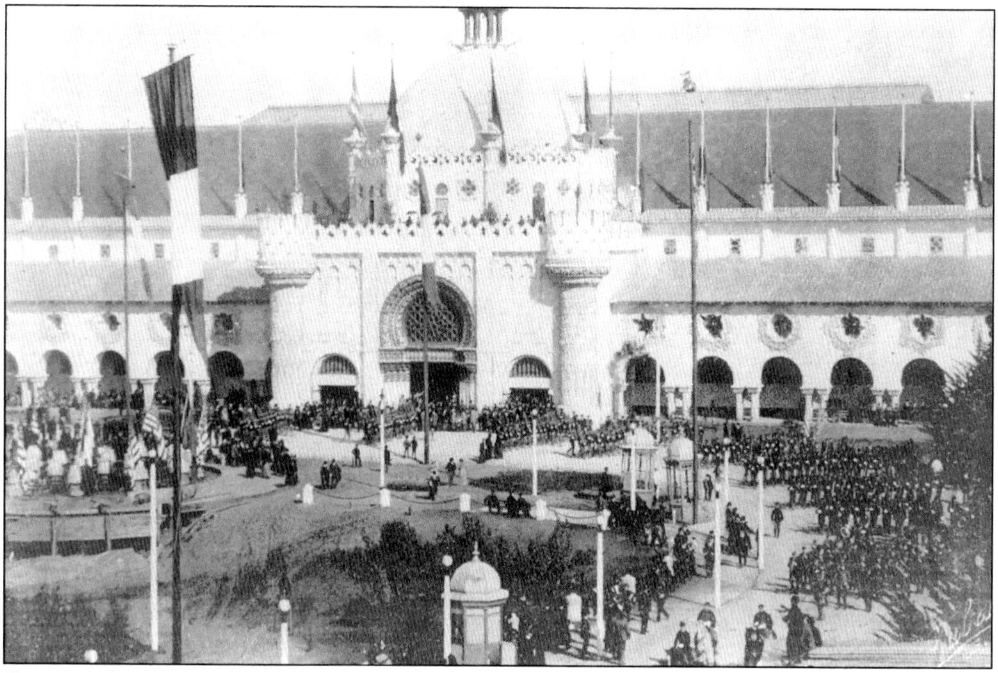

Opening day ceremonies naturally started with a parade. Units included four regiments of the California National Guard; one regiment of U.S. Infantry; two troops of cavalry; two batteries of artillery; the independent military companies of San Francisco and Oakland; representatives of civil, secret, and fraternal organizations; and very many marching bands.

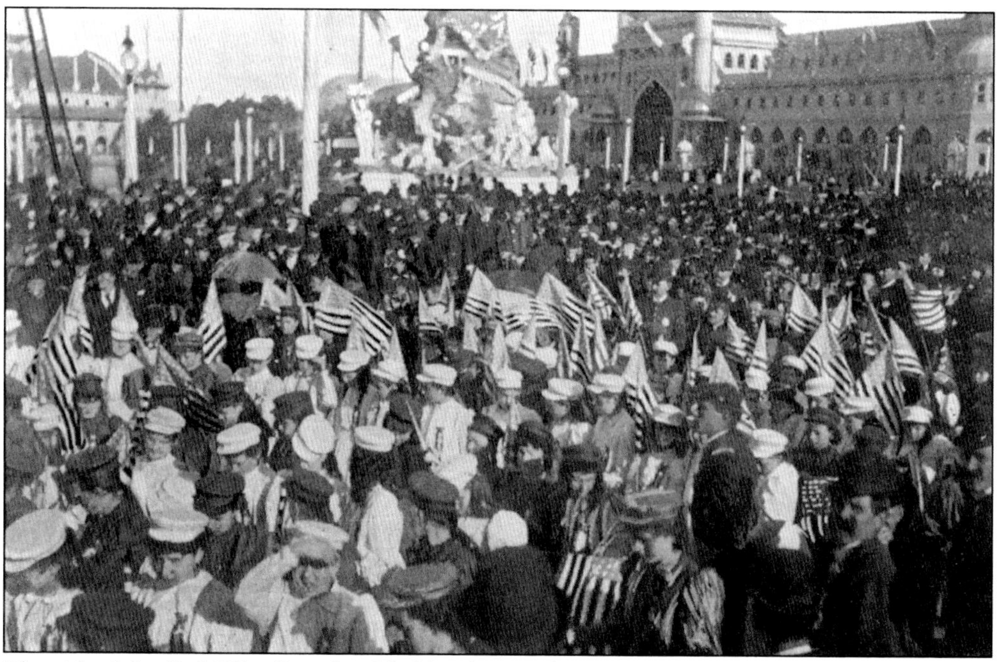

The girls of the Girls' Flag Brigade of Oakland not only carried American flags, they also dressed in red, white, and blue. The crush of people at the celebration became so great that at times it threatened to overwhelm many of the parade's contingents.

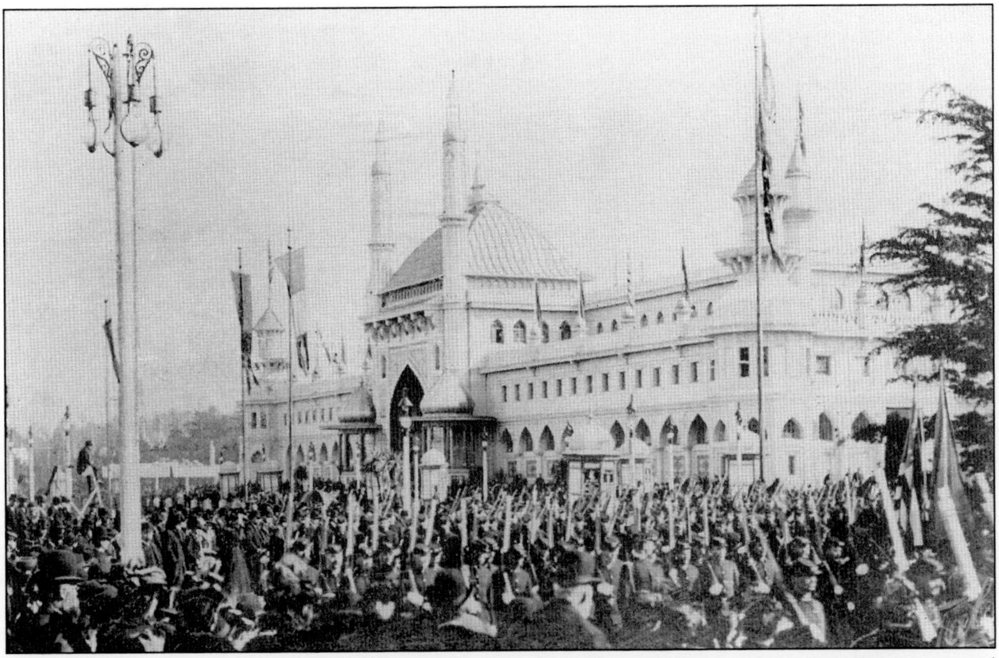

The opening day crowd was San Francisco's largest public gathering up to that date, a record broken twice more before the Fair ended. Many attributed the huge turnout to the mayor's proclamation closing all municipal offices, but the number of city employees then was fairly small. Perhaps his urging businesses to give employees the day off made the difference.

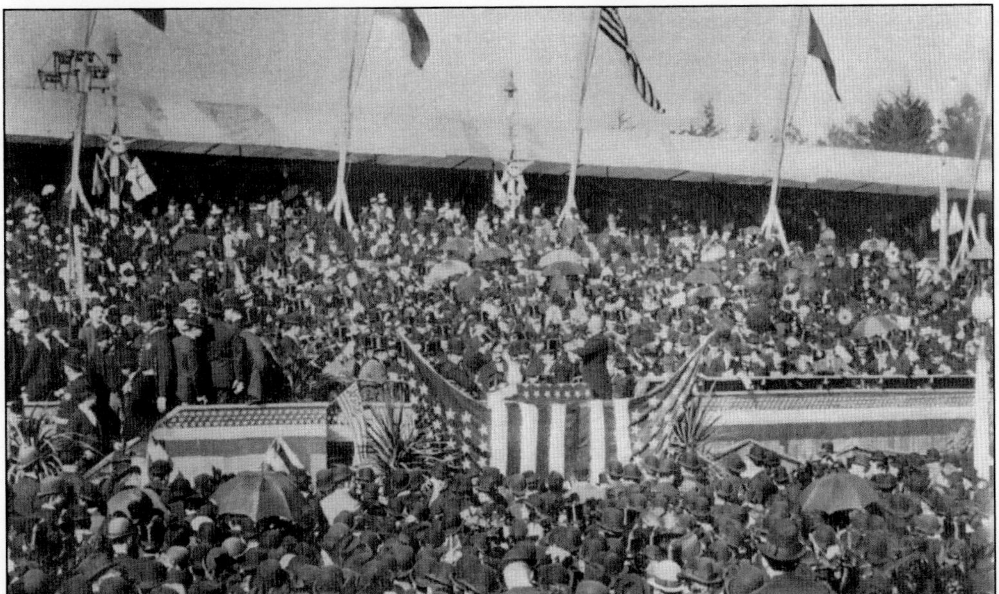

In those heavily formal times, hardly anyone except Director-General de Young, when he addressed them, was bare headed, a condition required for such a solemn occasion. "Join me in extending the hospitality for which we have a reputation to all strangers that come within our gates," he said. His wife pressed a button to start the Exposition's machinery and the Midwinter Fair by the Sunset Sea was officially open.

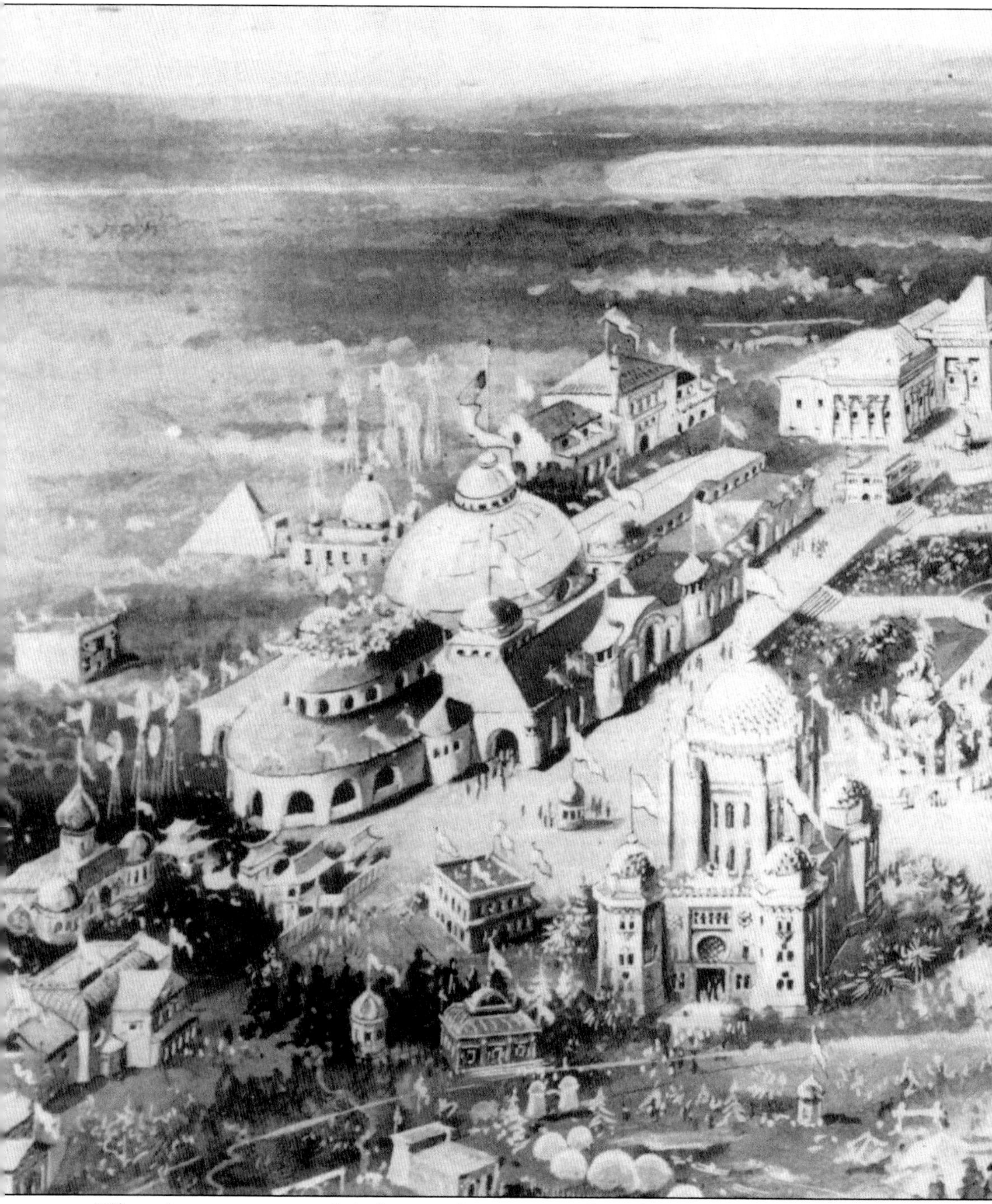

When the ceremonies ended, the crowd went everywhere, but especially toward the Midway, eager to see the celebrated attractions brought west from Chicago. Foreign nations didn't put up their own buildings, so goods were displayed by department, not by country, which helped both comparison and comfort. The Fair, a community unto itself, had its own police and fire departments, power plant, restaurants, theaters, hospital, and daily newspaper. More important, it had the sights and sounds, tastes and thrills of the world. Clockwise from the

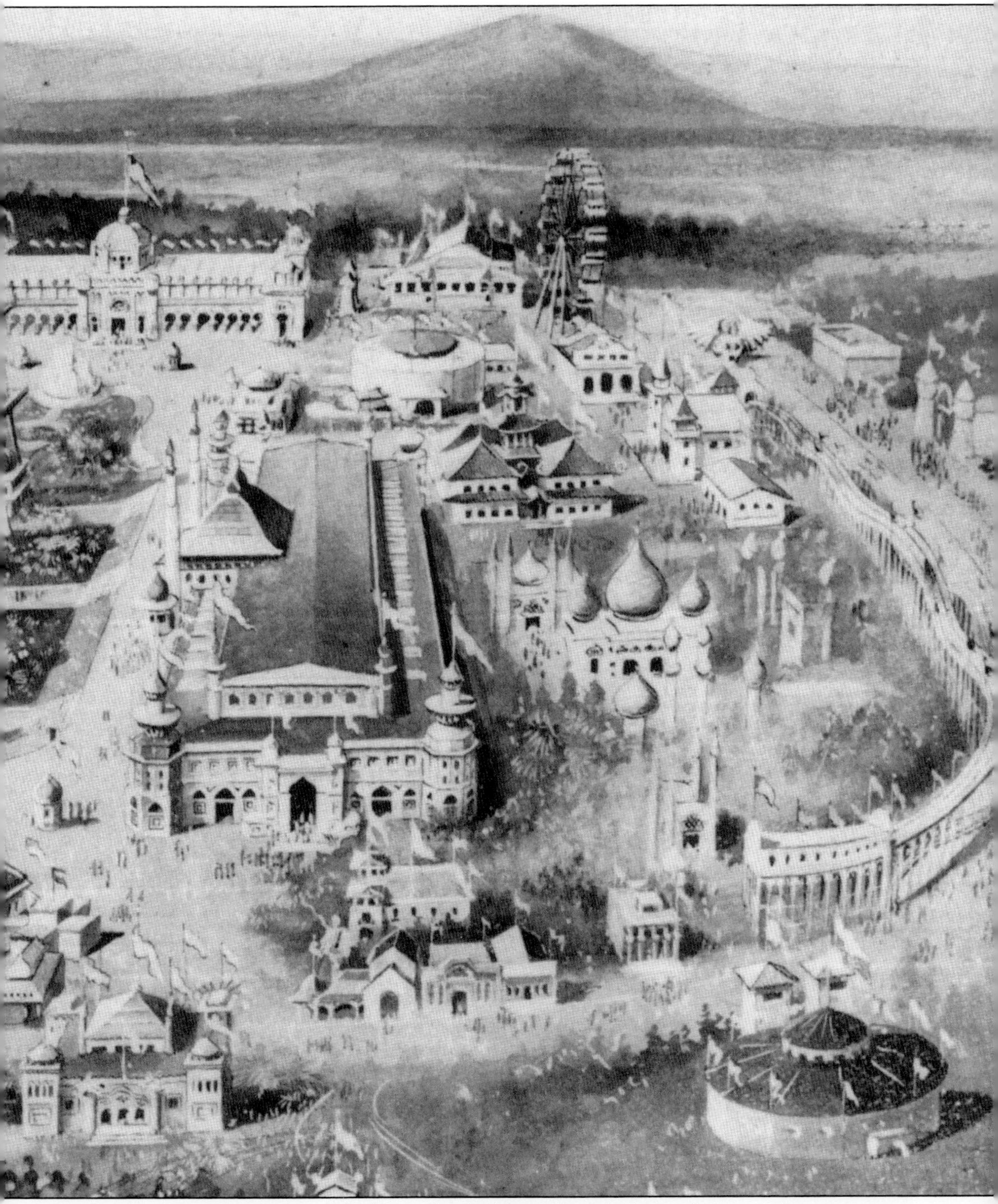

Manufacturers Building at 12 o'clock are the buildings for Mechanical Arts, Administration, Horticulture, and Fine Arts. Surrounding these are the attractions of the Midway: the Firth Wheel, the Santa Barbara Amphibia, the Cyclorama, the Chinese Pagoda, the Oriental Village, the Scenic Railway, Boone's Animal Arena, the Japanese Village, Anne Hathaway's Cottage, the Esquimaux Village, and much more.

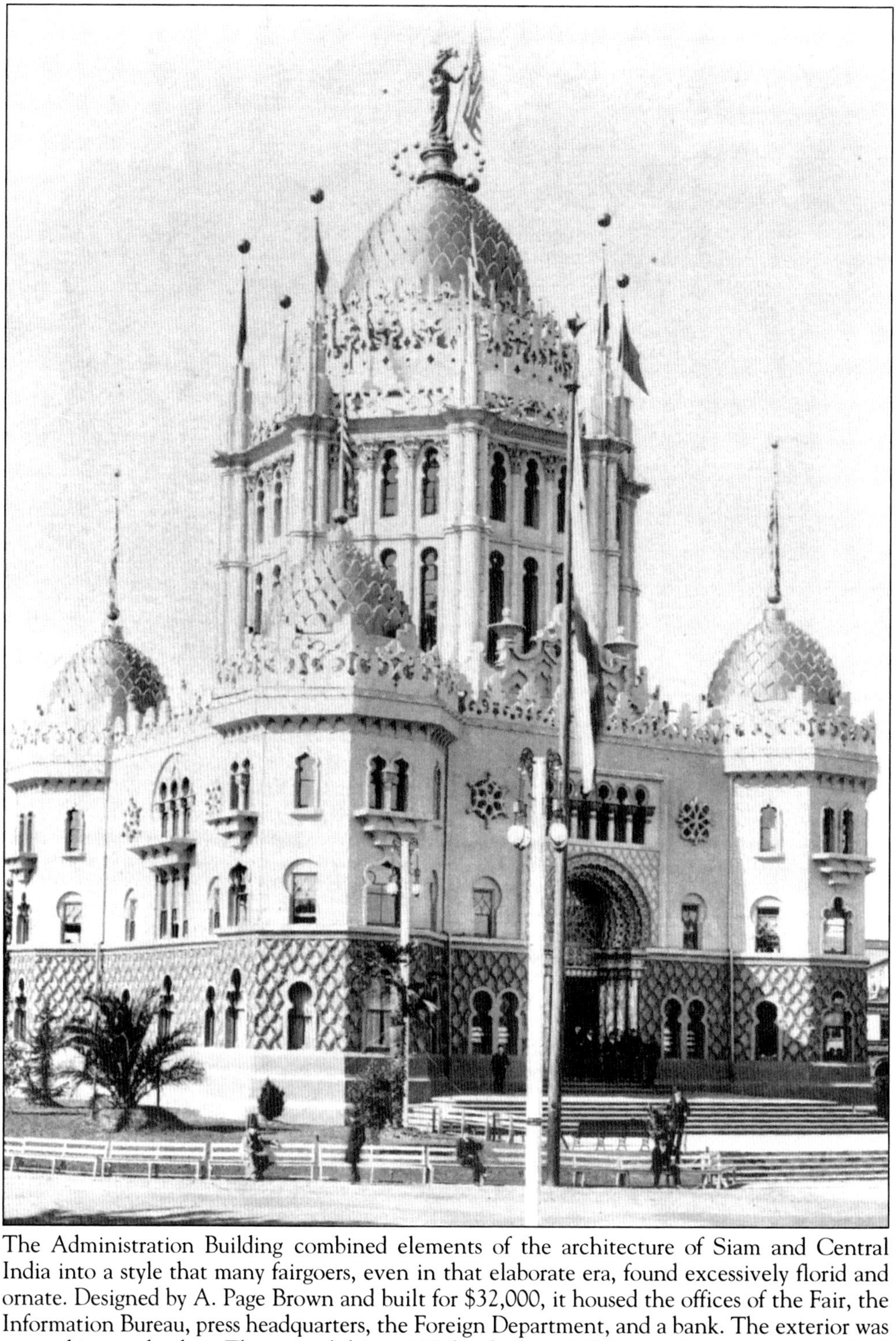

The Administration Building combined elements of the architecture of Siam and Central India into a style that many fairgoers, even in that elaborate era, found excessively florid and ornate. Designed by A. Page Brown and built for $32,000, it housed the offices of the Fair, the Information Bureau, press headquarters, the Foreign Department, and a bank. The exterior was painted in vivid colors. The central dome, 135 feet high and 50 feet in diameter, was finished with gold leaf. Fanciful murals graced the interior walls.

Two

PALACES AND PLEASURES

That it would be seen as a secondhand, shopworn version of the World's Columbian Exposition greatly concerned the Midwinter Fair's directors—and not without reason. Almost every foreign exhibit that came to San Francisco came from Chicago; so, too, most of the Midway attractions, along with the name Midway itself. Ticket booths, turnstiles, and exit gates—all had been built for the 1893 fair; the souvenir admission tickets were unabashed lookalikes. Even the Columbian and Midwinter Guards wore identical uniforms.

To create a distinct look, Director-General Michael de Young gave strict orders: the principal buildings of the Midwinter Fair should not copy the beaux arts, neoclassical style of the White City's exhibition halls. In response San Francisco's architects designed structures for the Court of Honor—another name first used in Chicago—that used a miscellany of styles from different places and periods to evoke California's architectural past. The result: an assortment of vibrantly colored national, state, county, and commercial buildings, stalls, stands, kiosks, gazebos and concessions, juxtaposed with no apparent plan or purpose, embellished with an archive of architectural elements and ornaments.

Did fairgoers like this mingle-mangle? *Overland Monthly* did. Wrote its critic, "No element of utility has been sacrificed to those of beauty, or symmetry, or color. Blue domes, dark red tiles, pearl-white marbles, and time-stained yellows, are all in harmony with sea and sky and earth and trees. The picture thus presented is, in a great degree, flawless."

Others were not so sure. One somewhat sour critic wrote that the Exposition displayed both incoherent intent and dismal design. Willis Polk, who became the Panama-Pacific Exposition's supervising architect, considered it entirely disgraceful.

Inside, the structures resembled huge warehouses. Except for the Administration Building, which housed offices only, each exhibit hall contained a different department of the Exposition: agriculture, machinery, manufacture, fine arts. Displays of a fifth department—isolated and collective exhibits—were dispersed throughout the fairgrounds.

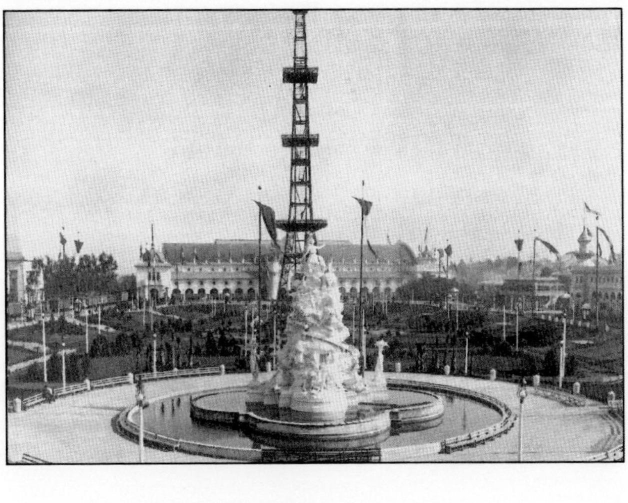

The Court of Honor was the Exposition's heart. After filling and grading Concert Valley, the system of terraces and steps built there to link the main exhibition halls became the true unifying factor in the Fair's grand design. When it closed, almost everything from the Fair was removed, but the Court of Honor—now the Grand Concourse—remained. Although somewhat changed, Golden Gate Park's museums stand around it now.

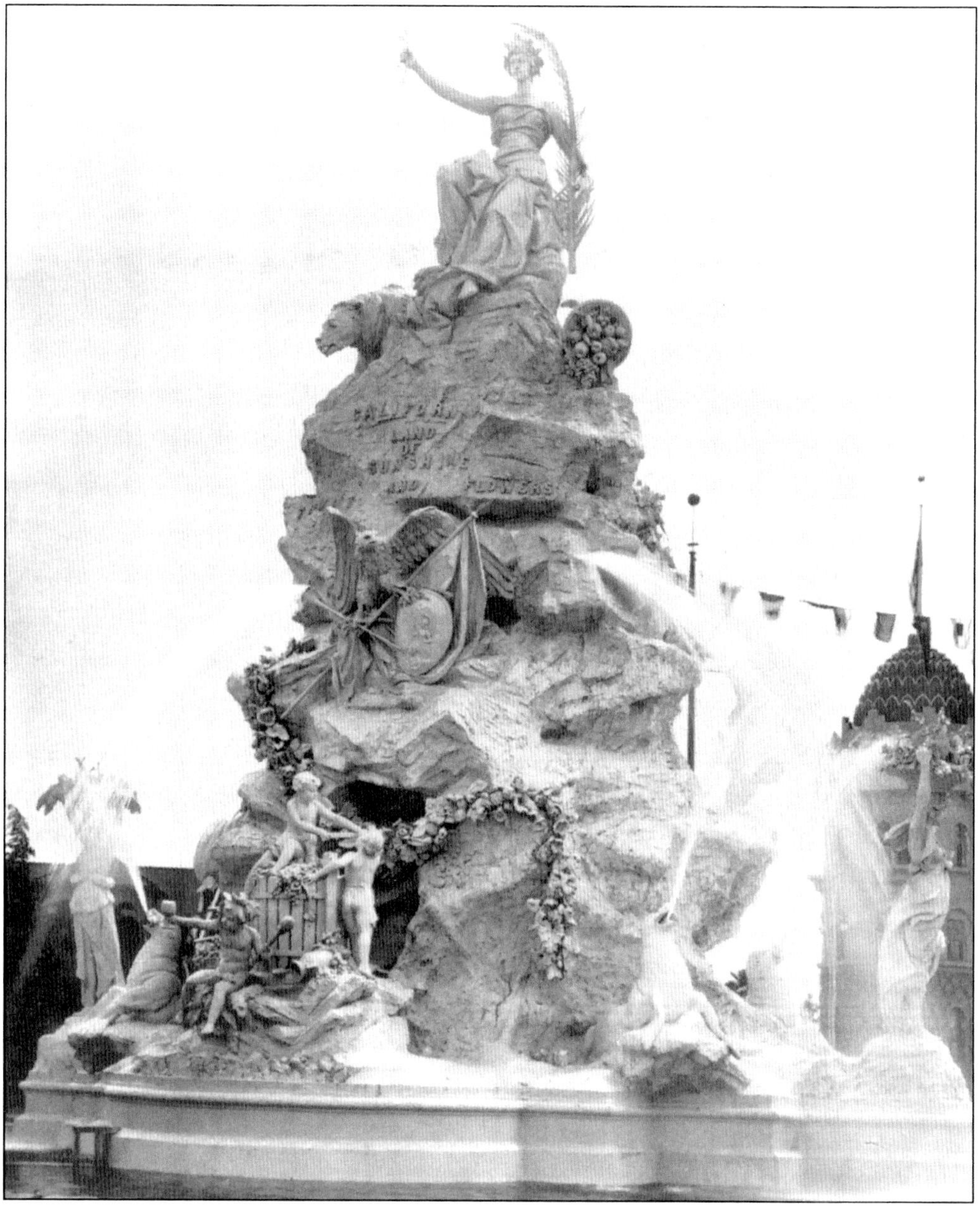

The California Fountain stood near the Administration Building on the western end of Court of Honor. An allegory in plaster and staff, it told the story of the state's abundant natural wealth using every possible symbol. Minerva, crowned with a wreath of golden poppies, sits upon her mountaintops, the guardian bear at her side, the horn of plenty at her feet. Beneath them the motto, "California Land of Sunshine, Fruit, and Flowers," states the Fair's theme. An American eagle clutching the Great Seal of the state signifies California's loyalty to the Union. Representing nature's bounty, Pomona, the Roman goddess of garden fruits, stands opposite Flora, the Roman goddess of flowers, vines, youth, and spring. Working a press, cherubs, who seem too young to drink legally, embody the wine industry. Miners, tools at hand, exemplify the state's mineral wealth. Seals denote the riches of the Sunset Sea. Done in high Victorian monument style, the fountain was the work of Rupert Schmid and Marion Wells.

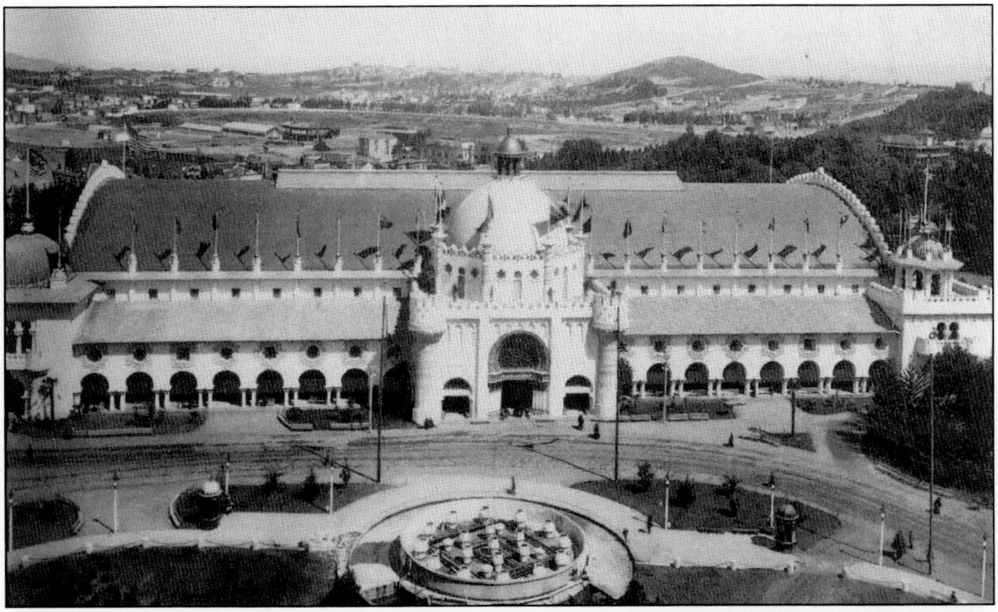

The Electric Fountain, foreground, in front of the Manufactures Building, was on the opposite side of the Court of Honor from the California Fountain. Simplicity itself, it was completely free of divinities and devices. By day a reflection pool, it became at night a brilliant aquatic display, with jet of water as high as 90 feet and colored lights playing on them from beneath.

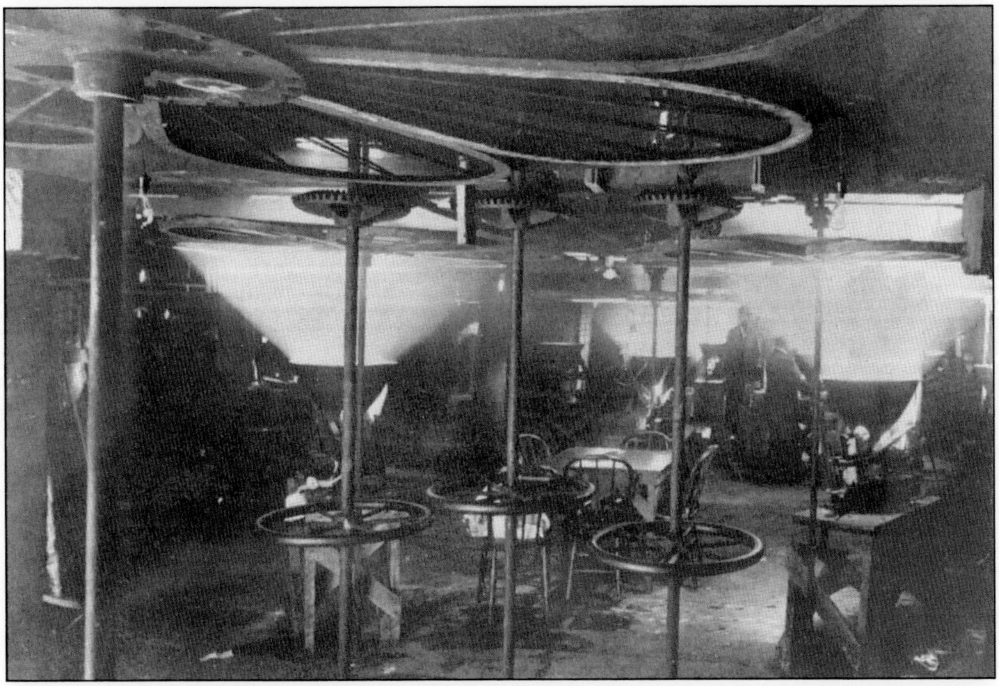

The room under the Electric Fountain housed all the equipment needed to create the display. Revolving panes of colored glass, lit from below by powerful incandescent lamps, produced the effects. The operator couldn't see the fountain, so an attendant above ground used a control panel that signaled him to make the water higher or lower, the lights now blue, now red.

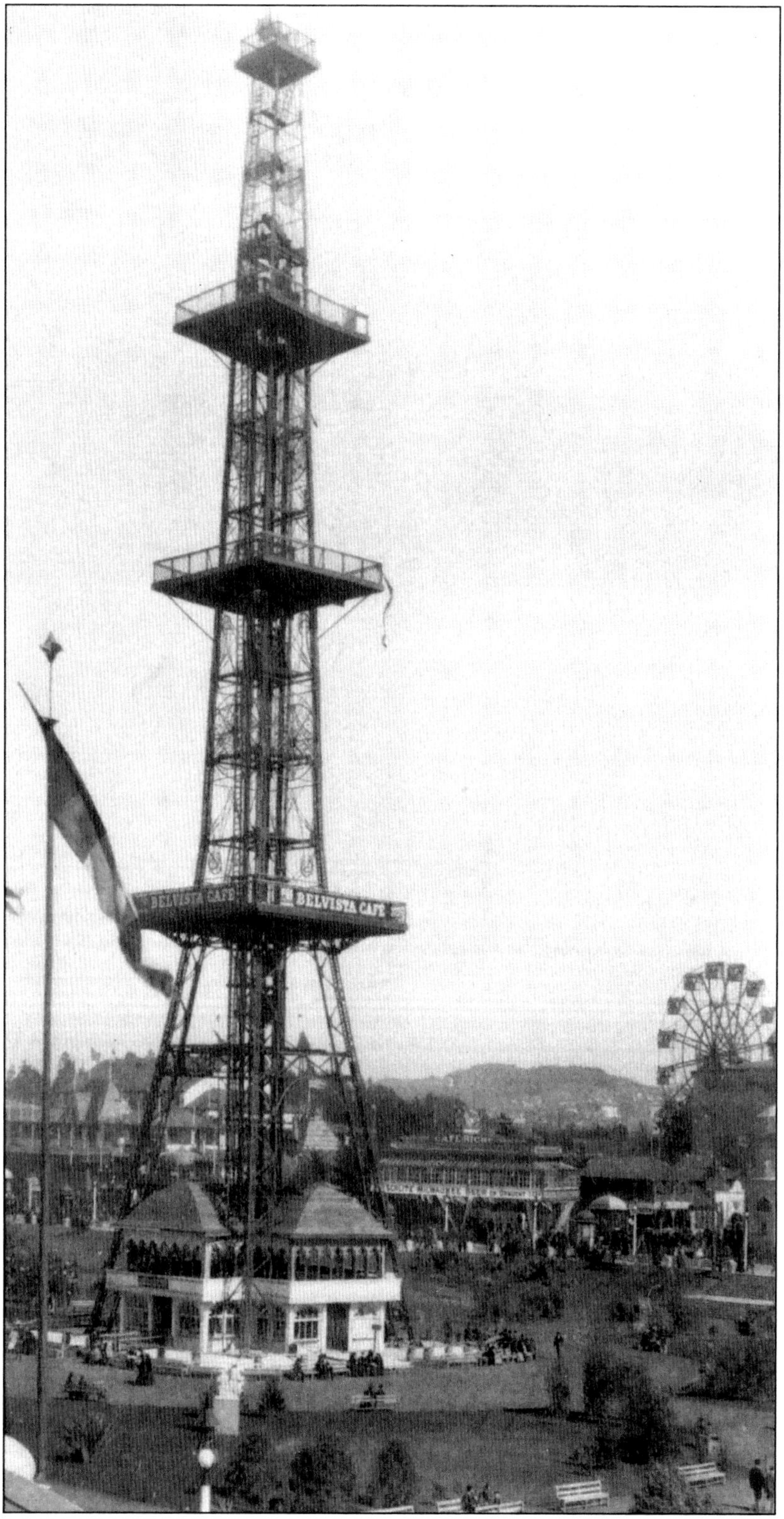

Bonet's Electric Tower stood at the center of the Court of Honor, *la petite Tour Eiffel de San Francisco.* At 272 feet it was the tallest structure on the fairgrounds. An electric elevator took passengers up to galleries at 91, 147, and 210 feet; the trip took about three minutes. The first landing of the Tower was home to the Belvista Café; fairgoers, too tuckered to continue after climbing to this level, could renew themselves with light refreshments or a lavish meal—or simply enjoy the panorama for once inside a stationary attraction. From the third landing, a spiral stairway led to the searchlight at the top, but it wasn't open to the public. Thirty tons of buried concrete anchored it all to the park.

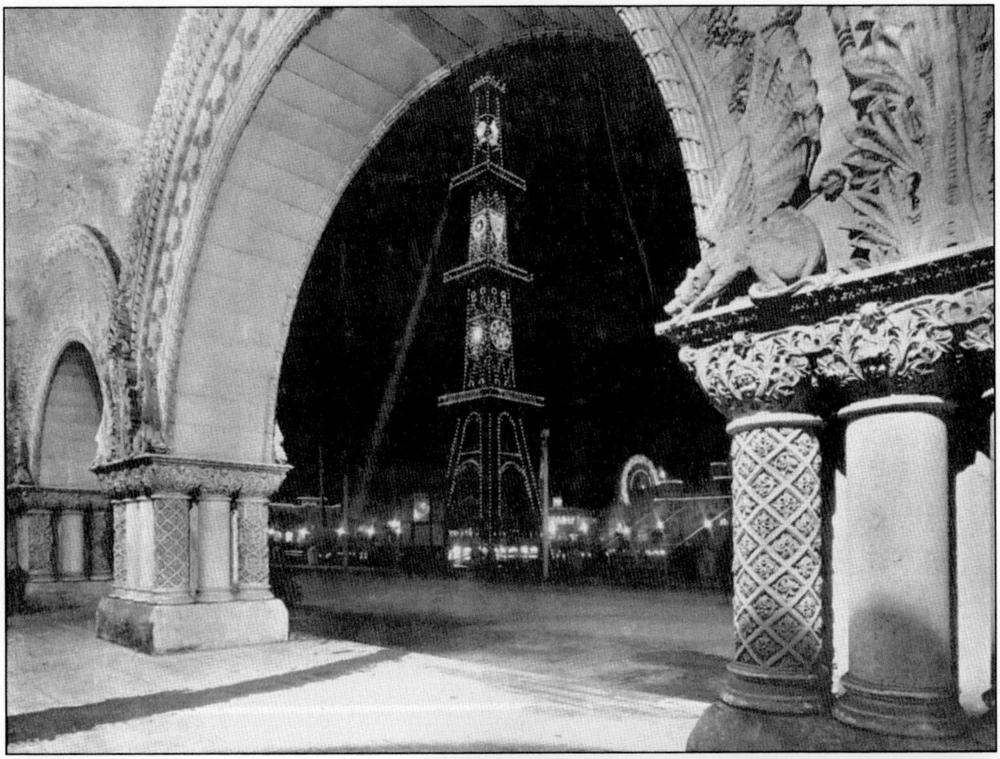

The Electric Tower at night, its searchlight illuminating the Midway, stood out vividly against its surroundings. It was a landmark that could be seen from every part of the fairgrounds. Here it is seen from the portico of the Agriculture Building. All the decorations of the archway were carved in plaster and staff, including the griffin on the right, poised resolutely to relish the next passerby.

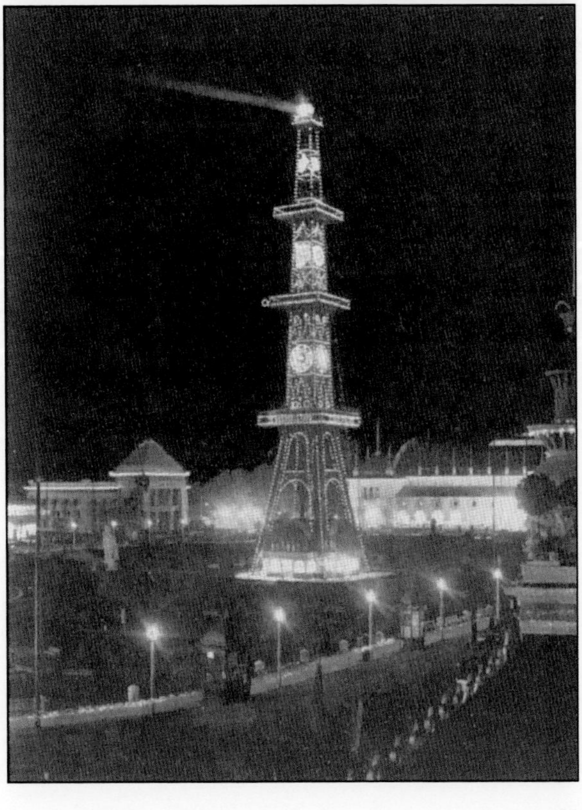

The searchlight at the top of the Electric Tower was the largest and most powerful then in existence, two facts the management publicized eagerly. On clear nights, its beam could be seen for miles in all directions.

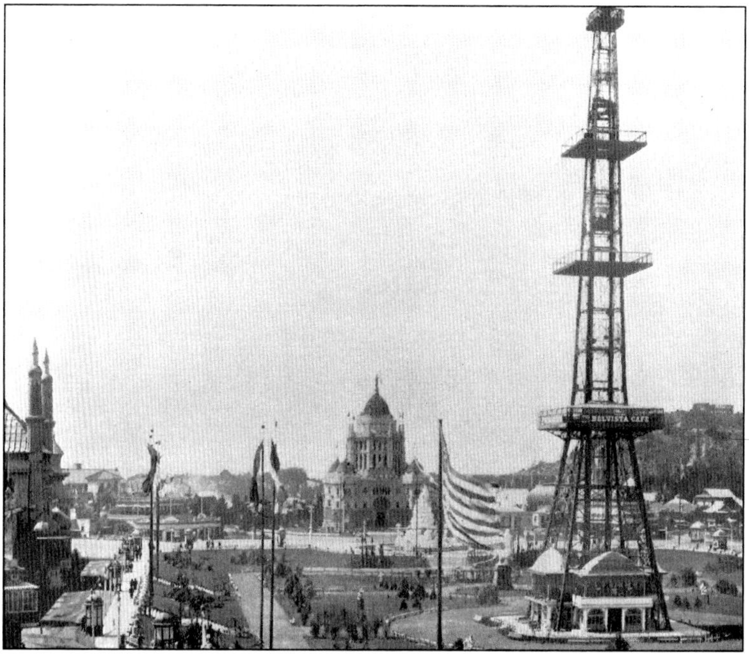

The compact size of the Exposition enabled visitors to browse its principal pavilions without very much walking between them. After all, they were either next door to each other or across the Court of Honor. Investigating everything inside them was another matter entirely, but strength permitting, it took less time and wear than it had in Chicago, for which, no doubt, many were thankful as the day progressed.

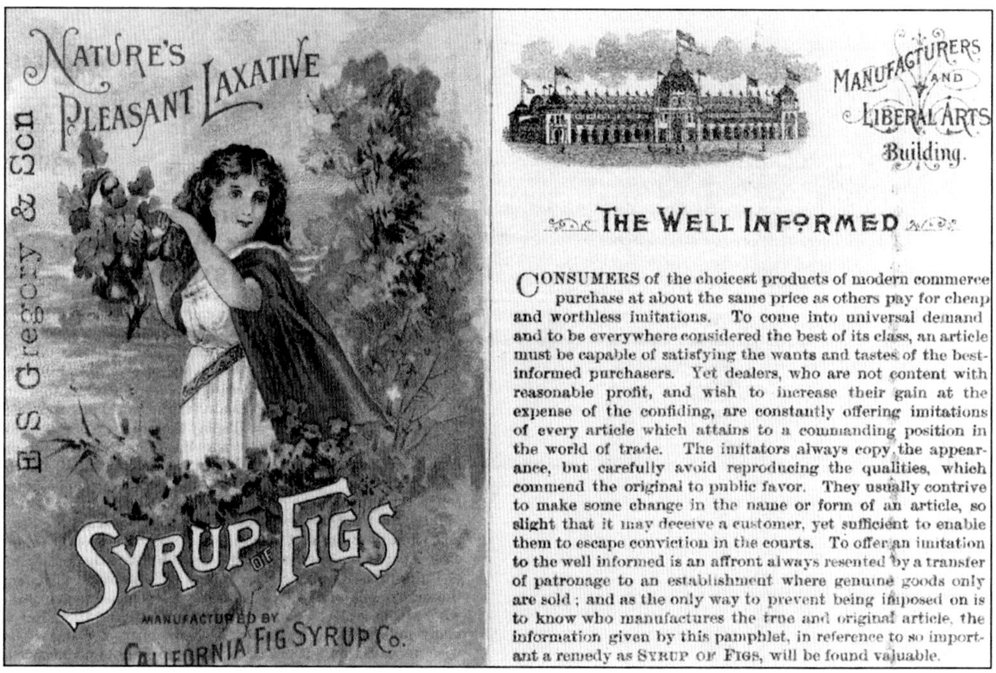

The Fair charged no admission to the four exhibit halls. Visitors were welcome to wander among the filigree and furbelows for as long as feet and family allowed. Stalwart guests with strong legs and large satchels could gather innumerable trade cards, circulars, brochures, booklets, broadsides, and samples, most of which, sadly, fell victim to a misguided house cleaning or two. Syrup of Figs' giveaway illustrated each of the main buildings.

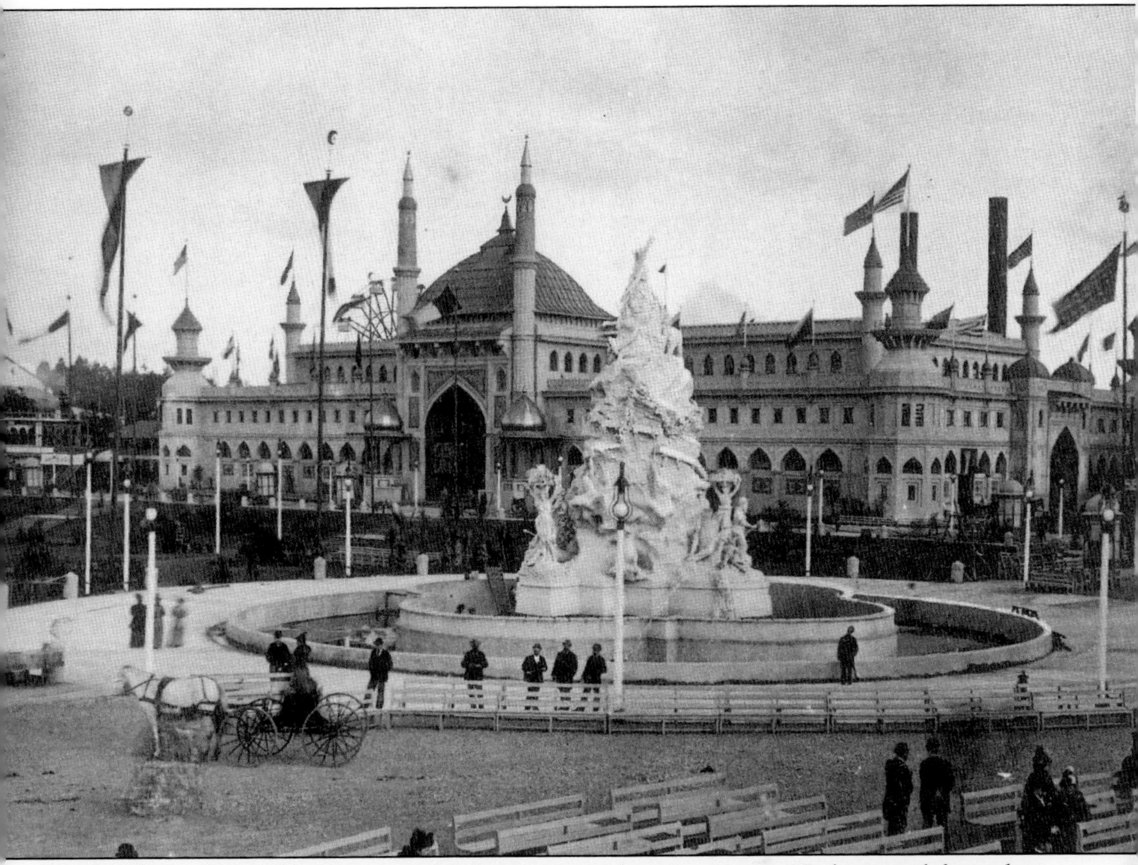

The Mechanical Arts Building was the Fair's second largest structure. It borrowed from the architecture of India and was "as purely East Indian in style as is possible to design a building of this class." Outside, the walls were painted ivory; the main entrance tower and spires were turquoise and yellow; the columns vermilion; the pinnacles tipped with gold. Inside, visitors saw displays of mining machinery, especially the newest equipment for digging, tunneling, shafting, and quarrying. They also saw machines that produced goods of all kinds made by mechanical equipment. The mineral exhibits, too, had to do mostly with mines and mining. The building also housed the Fair's bakery and the printing plant for the Exposition's daily newspaper and other publications. Architect Edward R. Swain made no attempt to harmonize the smokestacks with the minarets.

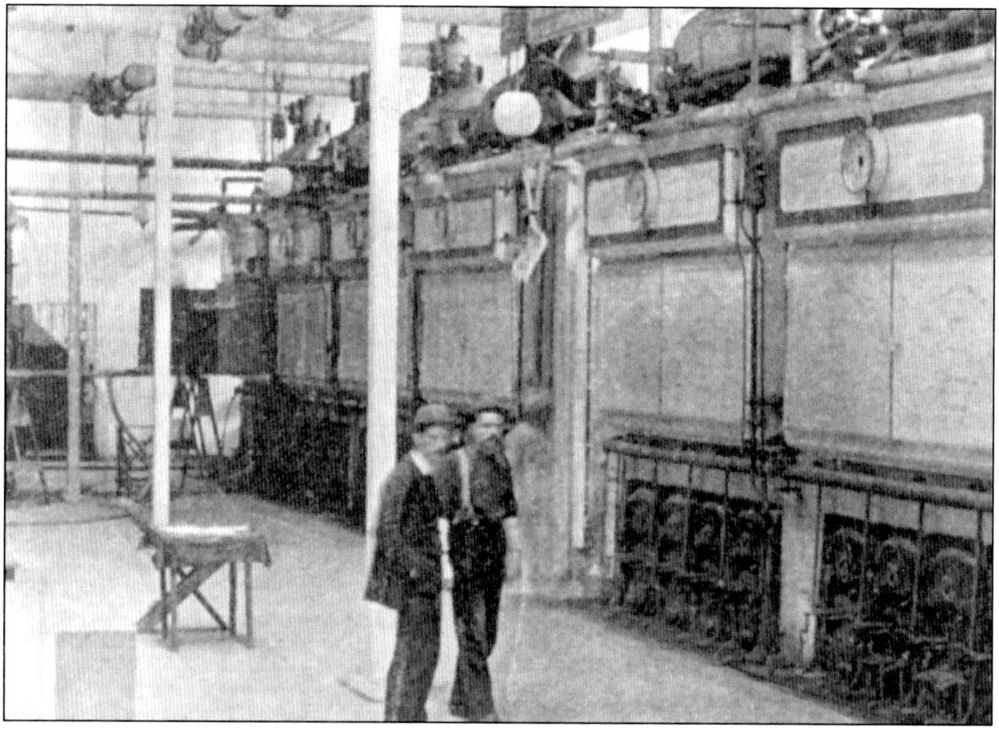

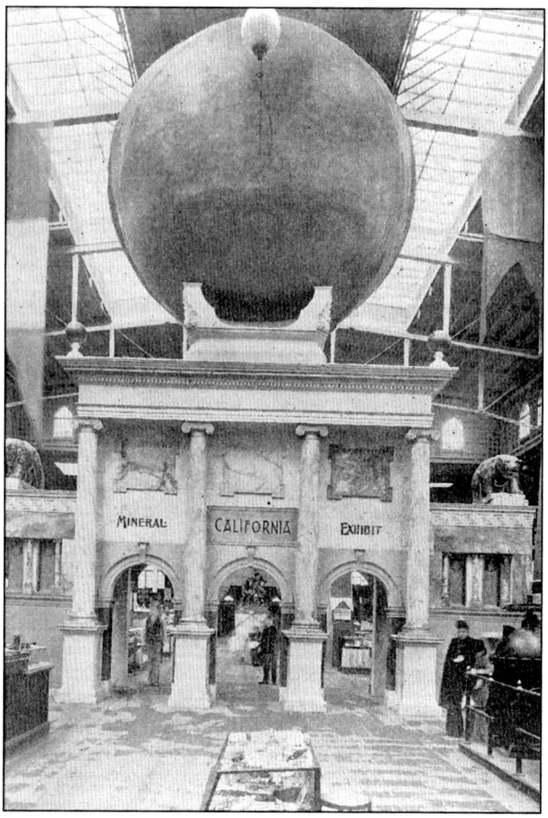

Eight Heine boilers supplied the steam to operate the engines on display. Because they were fueled with petroleum, not coal, they became an exhibit in their own right, the boiler room maintaining a "cleanly look." The assistants were so delighted, they hung their coal shovels on the walls tied with ribbons, each labeled, "At Rest." One of these may be seen between the second and third boilers from the right.

California's Mineral Arch separated the mineral displays from the machinery exhibits. It had been constructed for the Columbian Exposition, where it won a second place award. It looked like a solid stone structure, but in fact was a wood framework with a stone veneer. The ball represented all the gold mined in California, estimated to be 2,071 tons worth $1,248,272,935—this when gold was valued at $20 an ounce.

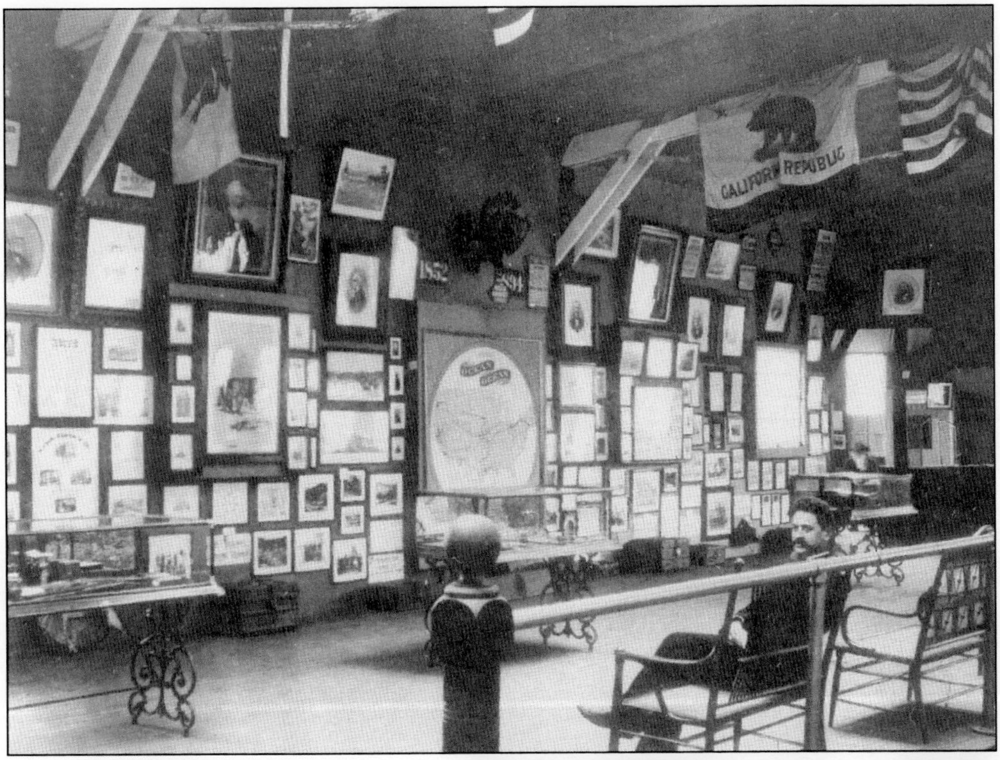

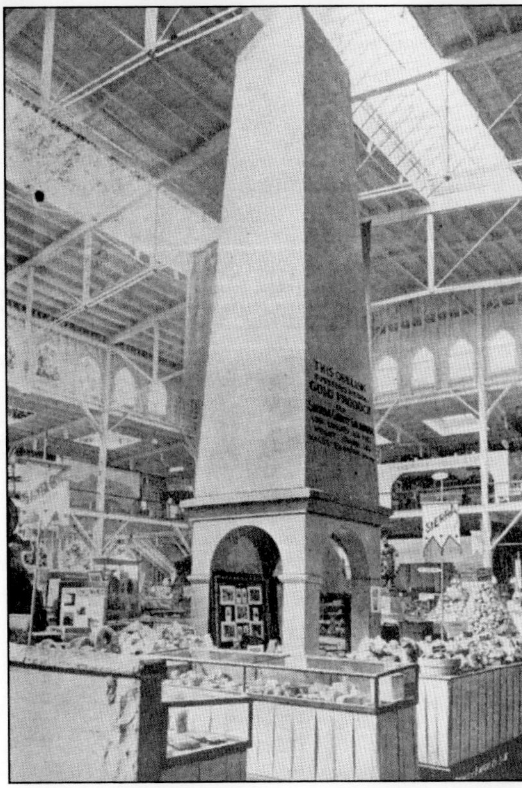

The Wells Fargo exhibit fascinated fairgoers with display of gold scales, treasure boxes, counterfeit gold ingots, and souvenirs of daring stage robberies. Among others, these included a valise used by "Black Bart the PO8," who robbed 28 stages before his capture, and the pistols used by the Ruggles brothers in a stickup only two years earlier. The company also maintained an express station on the fairgrounds.

Sierra County's exhibit was typical of those found in the building. The obelisk represented all the gold mined in the county to date. Ore samples completed the display.

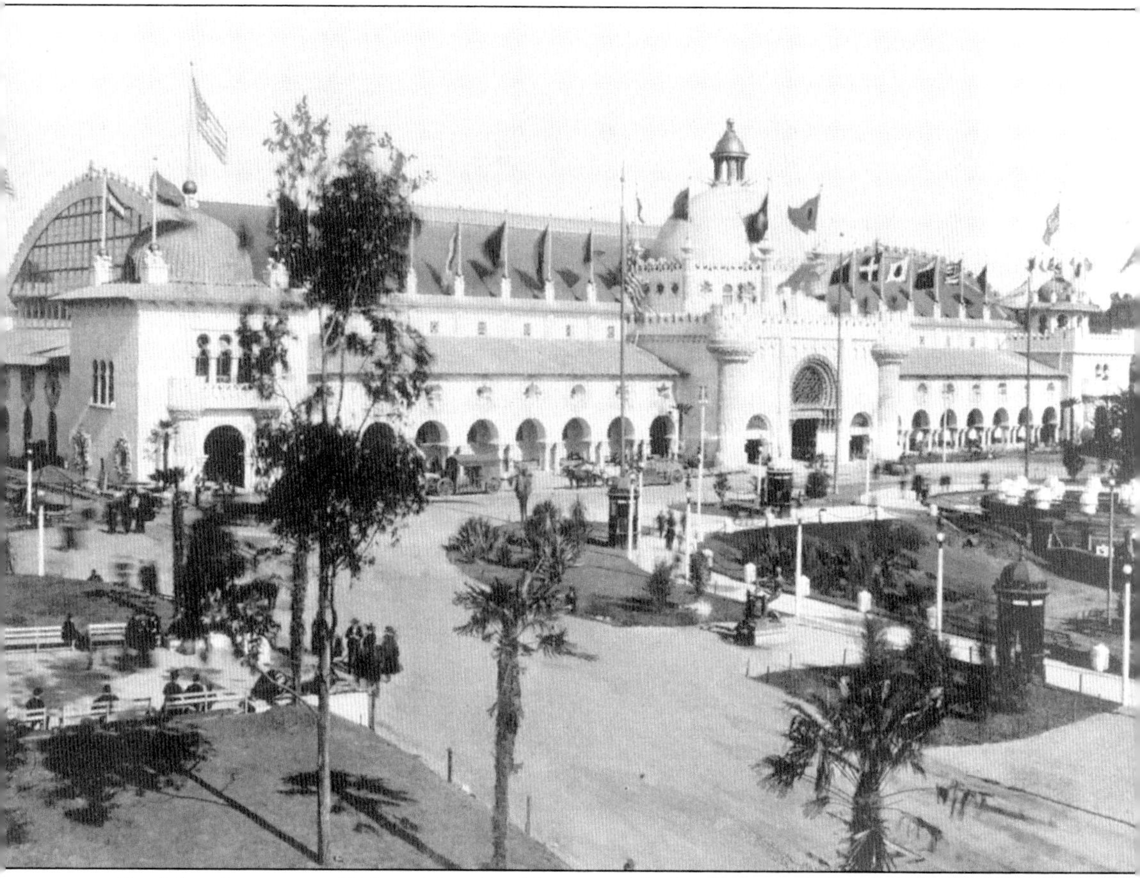

The Manufactures and Liberal Arts Building was 450 feet long, 200 feet wide and 92 feet high. It had 177,000 square feet of display space, making it the largest structure on the fairgrounds and the largest built in the West up to then. A. Page Brown, who also designed the Administration Building, created a facade that combined elements of Moorish and California mission styles into a fantasy of curled, red roofing-tiles; square, domed corner towers; richly ornamented trefoil and quatrefoil windows; and castellated turrets. The exterior walls were painted a cream color, and the central dome, 130 feet high, was painted turquoise, its cupola painted gold. The flags, colors, and coats of arms of exhibiting nations flew from the poles that encircled the roof of the building. In addition to its exhibits, the building also housed a carillon that played daily at 1:00 p.m. and 6:30 p.m. The completed structure cost $120,000.

Entering the Manufactures Building, visitors clearly saw the structure's Moorish origins. Soaring more than 60 feet, how could patrons not be impressed by its massive doors and towering archway? Passing beneath a central dome 130 feet high and 56 feet in diameter, they found an interior whose main trusses covered a span of 158 feet between its galleries.

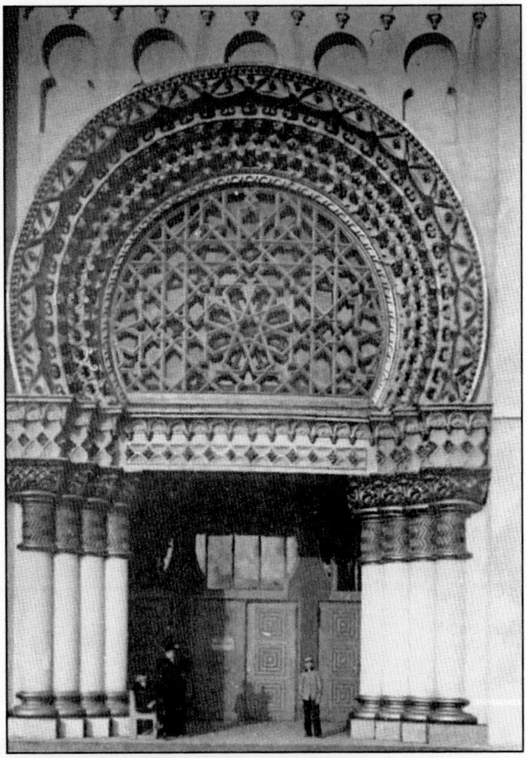

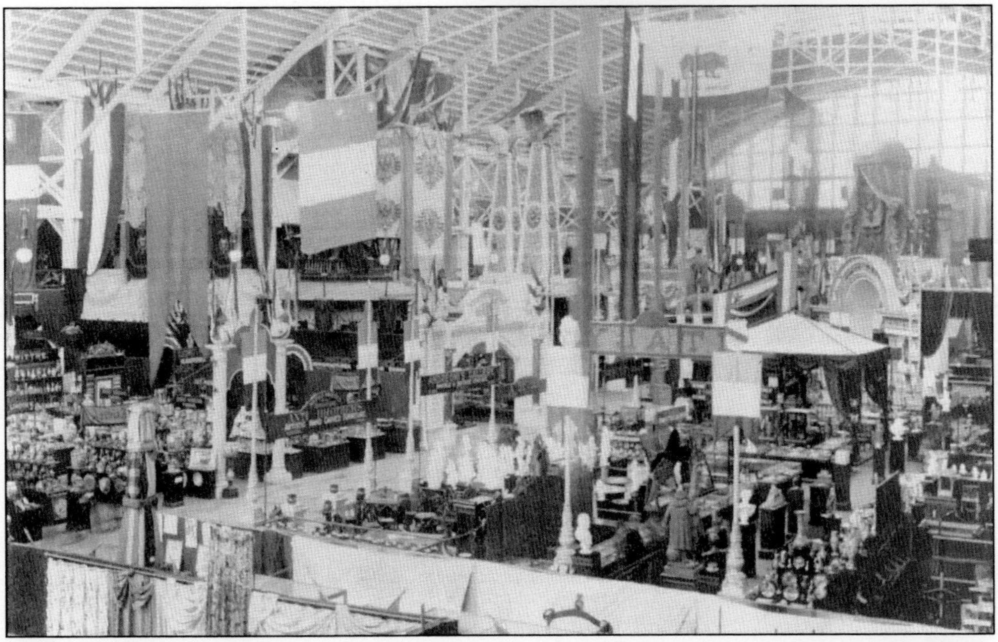

Inside the building fairgoers encountered displays from France, Austria, Japan, Italy, Great Britain, Russia, "the oriental countries," and many others. Those with strong constitutions and sensible shoes could compare the world's newfangled ideas in every enterprise from pottery to woodenware, tracery, needlework, optical goods, billiards, brocade, glass and crystal, harnesses, bicycles, soap, underwear, chocolate, embroidery, shoes, and more. Much of it was for sale.

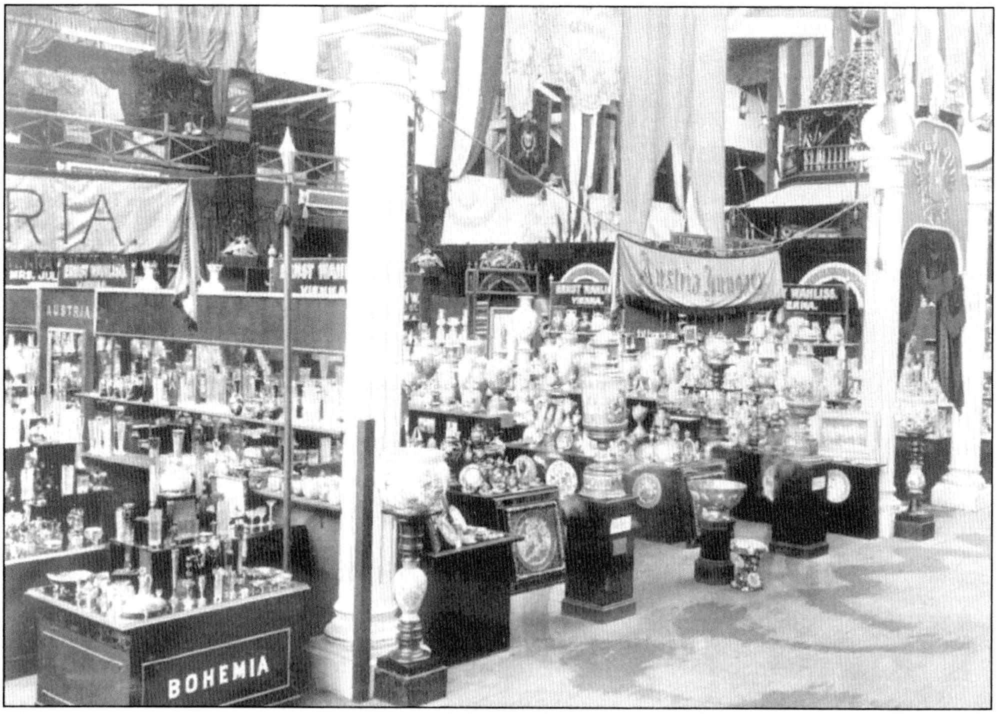

The Austria-Hungary booth was typical of the displays visitors saw. It showcased fine pottery and glassware in an abundance of forms and shapes. The decorative floor and table urns at the entrance, perfect for one's vestibule, anteroom, or salon, only previewed the quality of the bowls, vases, trays, plaques, and figurines inside.

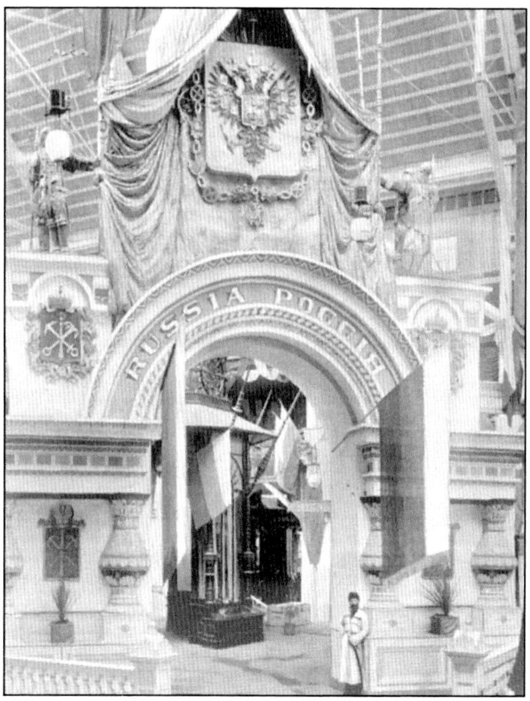

The Russian exhibit was one of the largest and most imposing at the Fair, beginning with the guard at the entrance. It showcased an extensive collection of furs. Also on display was an assortment of glassware, decorative arts, and fine tableware.

Bronze work from San Francisco's Pioneer Monument was displayed at the Fair before installation in the Civic Center, where it was dedicated on Thanksgiving Day, 1894; it resides there still. Here the figure of Minerva, taken from the Great Seal of California, towers above a woman with a very small booth and a very difficult task: selling Isabella commemorative quarters, left over from the Columbian Exposition for which they were minted. The coins cost $1 each. Most fairgoers were unwilling to pay that much for a 25¢ coin, so sales, slow in Chicago, were even slower in San Francisco. Who could imagine that a pristine example, if they wanted one, would cost their great-great-great grandchildren thousands of dollars? The gentleman on the left apparently has wandered over from the Russian exhibit behind them. From where the gentleman on the right has wandered is now unknown.

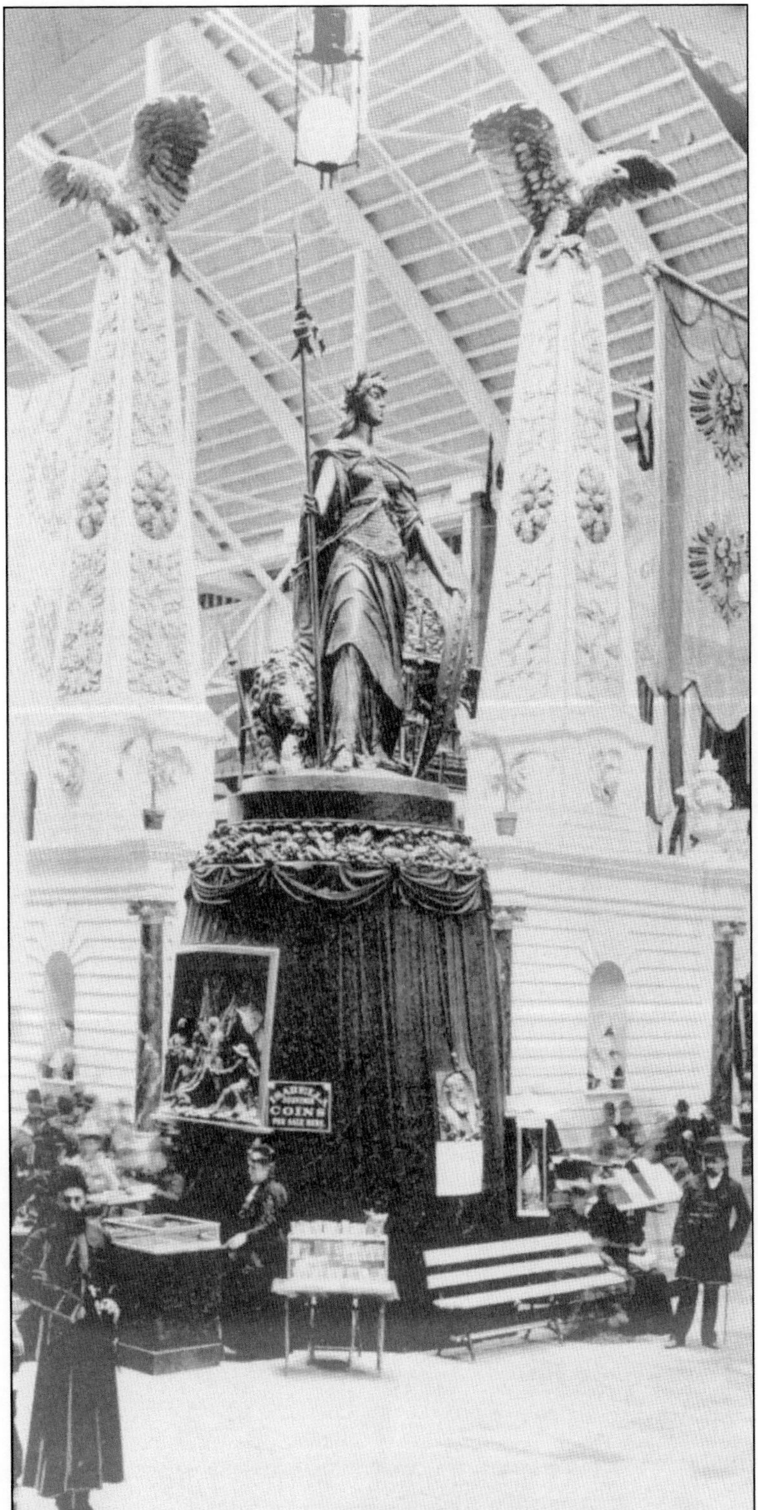

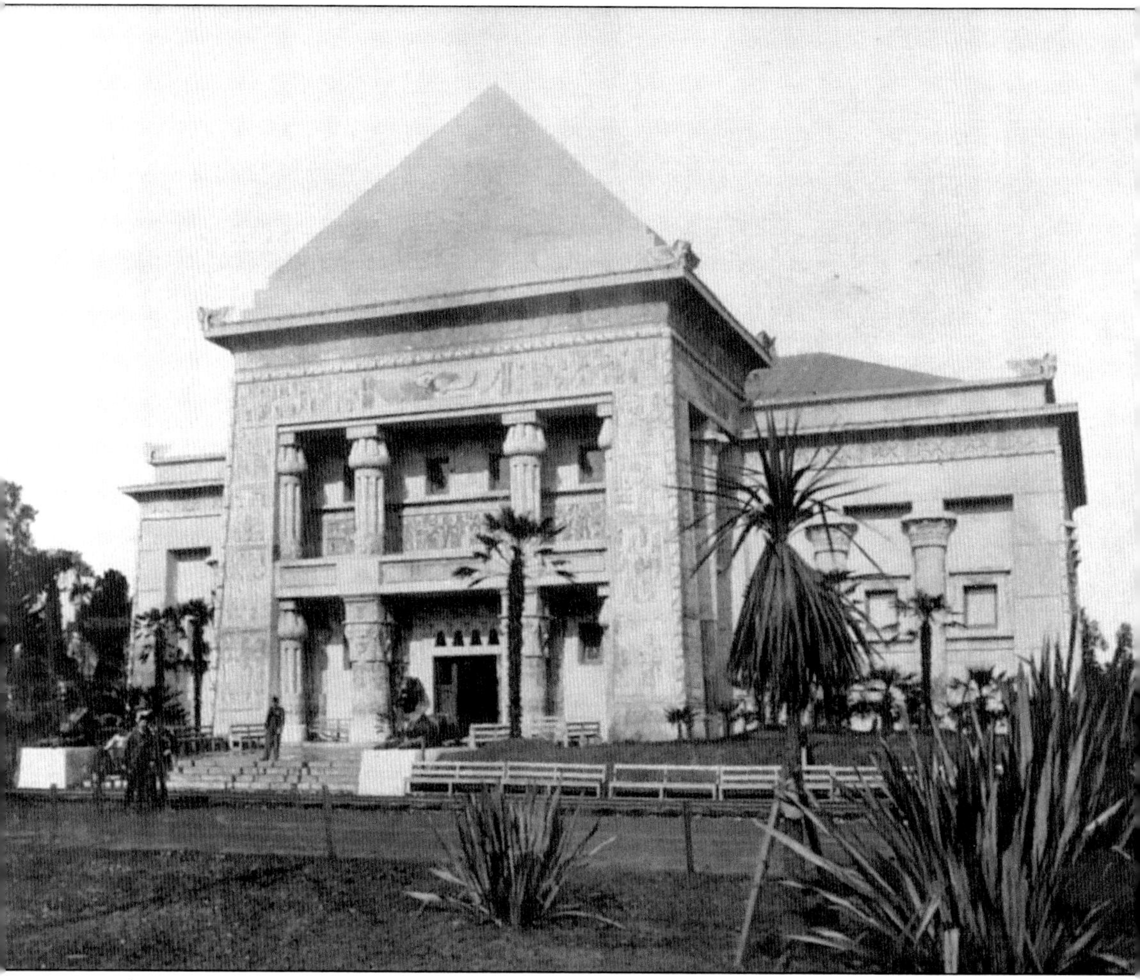

The Fine Arts Building was designed to invoke the repose and dignity of an ancient Egyptian temple, which at least one commentator thought was "befitting a building for the exhibition of real art." Palm trees and sphinxes added to the ambience, as did winged deities perched on each of the building's corners. Crossing the threshold, fairgoers passed replicas of columns found at the Temples of Karnak, Edfu, and Luxor. The frieze above the building's columns contained Egyptian sacred symbols and scenes of Egyptian life, all precise reproductions. The floor mosaic depicted traditional Egyptian figures. Hieroglyphics decorated the walls. Early renderings of the facade also included three-dimensional elephants at the bases of the second floor pillars and skylights in the pyramid roof, but neither remained in the final project. C.C. McDougall designed the building. It cost $60,000 to construct and was the only building meant to remain after the Fair closed.

Inside the Fine Arts Building visitors found an eclectic and thickly displayed collection of decorative arts, watercolors, sculptures, and oil paintings; the official catalog helped somewhat to sort it all out. Many of the artists are known now only to specialists, but the exhibit included work by Corot, Millet, Pisarro, Courbet, Sisley, and the first Renoir and Monet shown on the West Coast. No Latin American, African, or Australian artists exhibited.

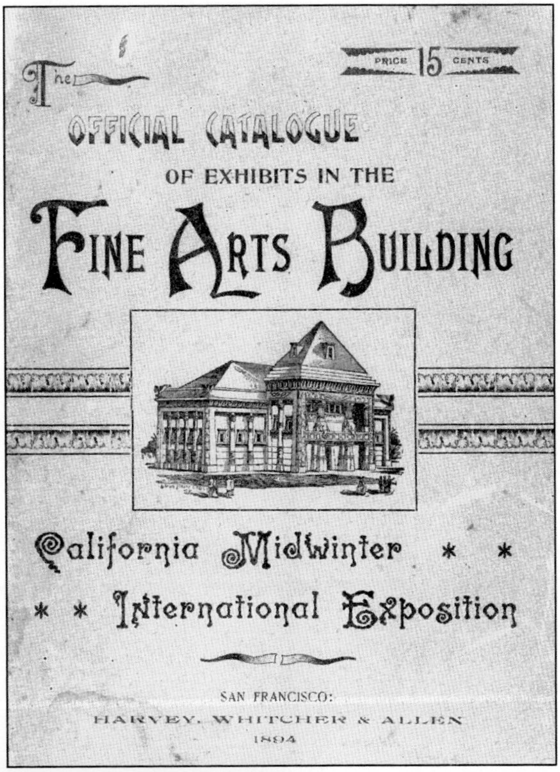
PRICE 15 CENTS

The Official Catalogue of Exhibits in the Fine Arts Building

California Midwinter * * * * International Exposition

SAN FRANCISCO:
HARVEY, WHITCHER & ALLEN
1894

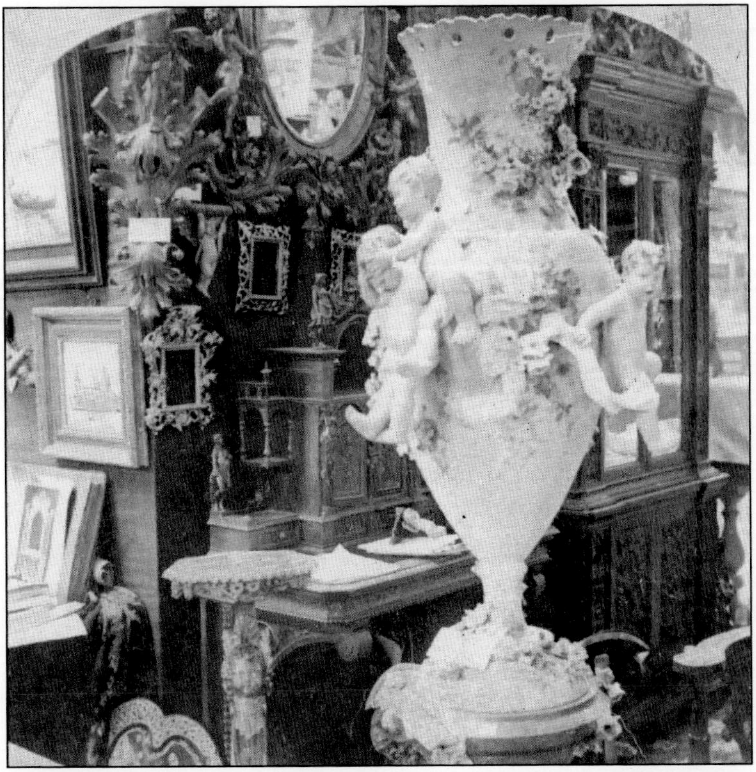

The Fine Arts Building's collection impressed local critics. How many younger students of the impassioned brush and the inspired chisel, alumni at last of art appreciation lessons, declined to set foot in the building again has not been tallied. The elegant and delicate vase shown here, a masterpiece of cherubic proportions then valued at $500,000, surely convinced many patrons of the value of great art.

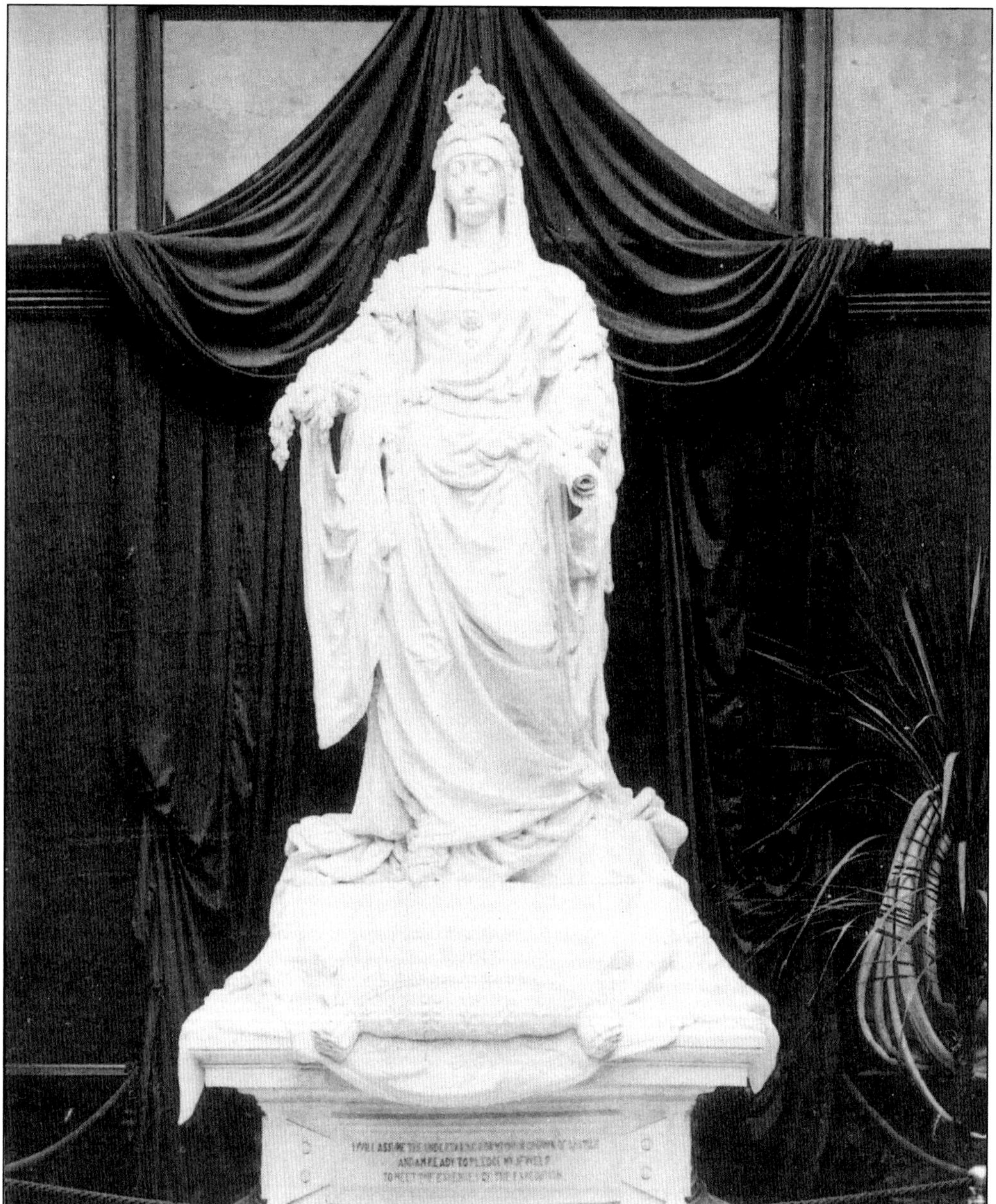

At the first floor gallery's far wall stood the *Statue of Isabella* by Harriet Hosmer, whom many considered to be the "greatest living woman sculptor." Standing 14 feet tall, including the base, it had been sculpted in plaster for a place of honor at the Columbian Exposition, which it didn't get. First it became the victim of a power struggle between the Isabella Association and the Fair's Board of Lady Managers. Then it was caught up in a controversy regarding Isabella's character and politics. Finally it was relegated to a spot in front of the California Building. Though the work was as controversial in San Francisco, it received a gold medal and consideration for the collection of the new municipal art museum. When the Fair closed, but before it could be purchased, the statue disappeared; it has never been recovered. Someday, perhaps, a spirited civil servant will creak open the door of a forgotten city warehouse and come face to face with the lost queen. A bronze casting went into the collection of Pope Leo XIII.

Suzuki's massive and beautiful cloisonné vase was a favorite of fairgoers. Including its base, it stood 8 feet tall. It was mounted in silver with silver handles; the pedestal was inlaid with carved panels. Before the Fair closed, it was purchased for the new municipal art museum. The image here is from a hand-tinted stereo card.

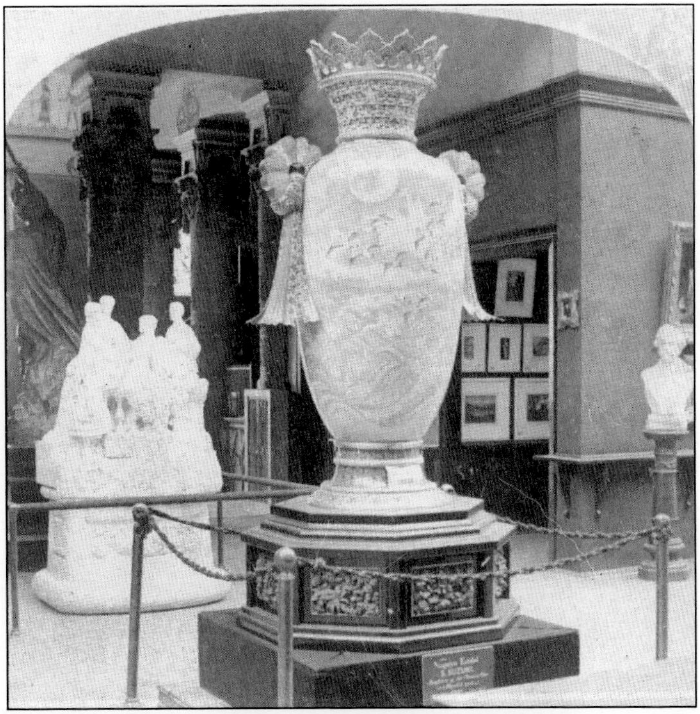

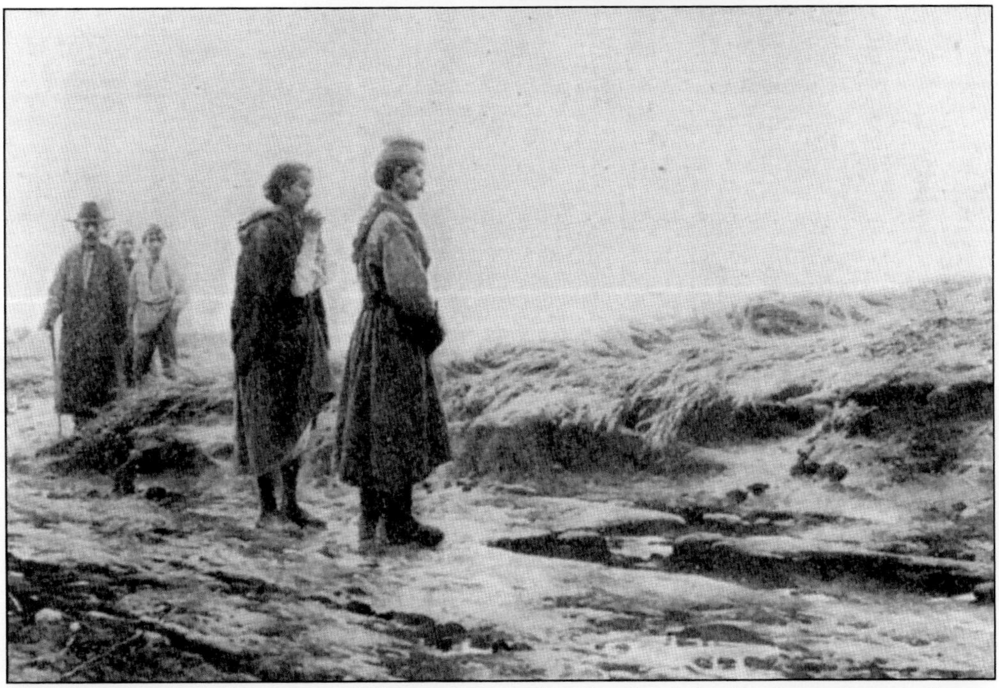

The oil paintings in the Annex, following the preferred style of the day, hung frame to frame, floor to ceiling; for novice art lovers, the effect could be somewhat intense. If the collection lacked quality, it didn't lack quantity—works by Polish artists alone filled three galleries. Typical was *After the Storm* by Thadeus Popiel of the Academy of Cracow in Austrian Poland, which won a gold medal.

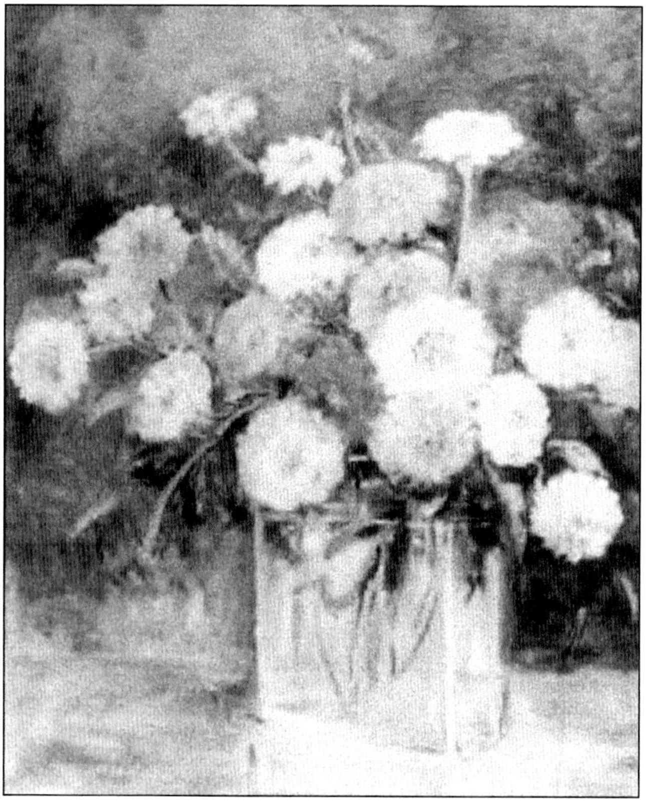

Women artists in addition to Harriet Hosmer displayed oils, watercolors, and drawings. A strong collection of Californians featured works by Clara McChesney, shown here, and Alice B. Chittenden, Dora Williams, Evelyn McCormick, Matilda Lotz, Eva A. Withrow, and many others. If nothing else, the exhibition showed that women, when given the opportunity, could produce works of art as outstanding and as mediocre as their male counterparts.

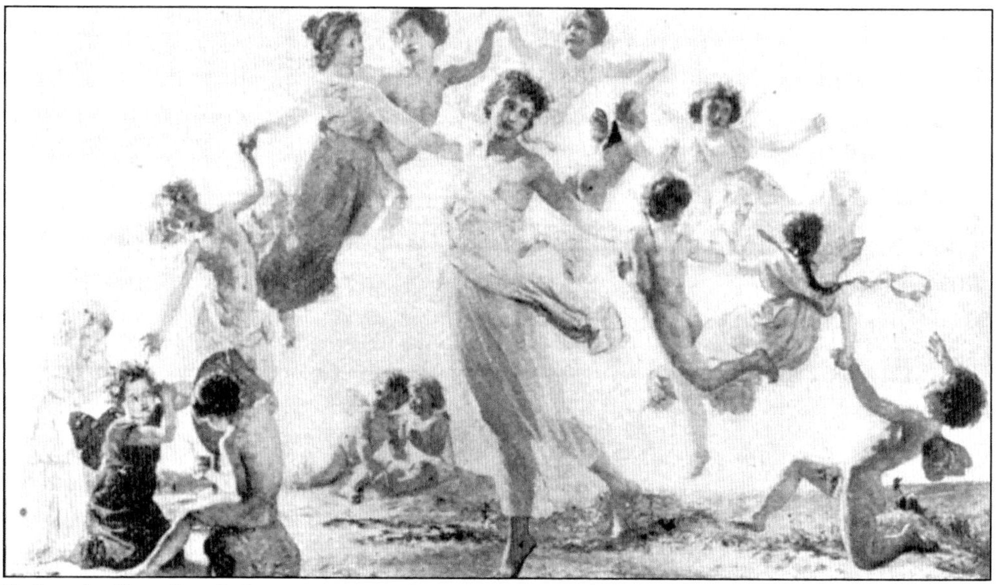

Before the Fair closed, there was some discussion to purchase the entire collection of paintings, including *Dance* by Adalbert Gerson, for the new museum, but it didn't happen. The Fine Arts Building reopened on March 25, 1895, as the Memorial Art Museum, the city's first municipal art gallery, with a miscellany of 6,000 items from the Exposition, de Young's private art collection, and thousands of Gold Rush artifacts.

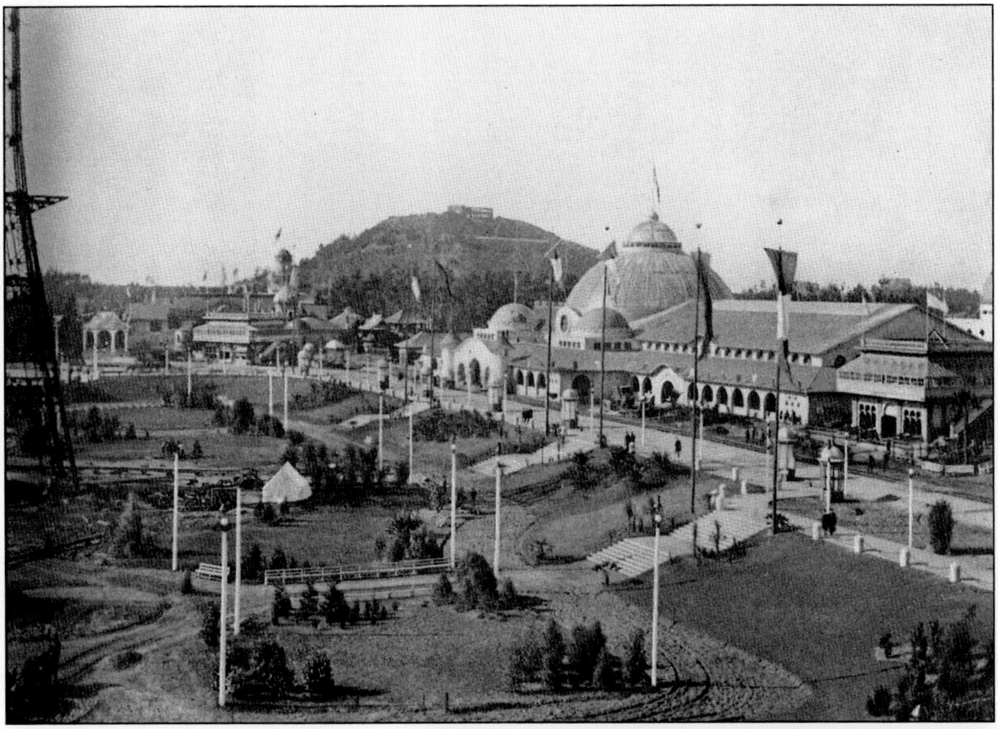

The Horticulture and Agriculture Building, designed by Samuel Newsom, was home to displays of fruits, grain, shrubs, flowers, and the implements to plant, cultivate, and harvest them. It combined California Mission architecture and Spanish Romanesque detail into a faux-adobe fantasy unlike any other. The structure was 400 feet long, 190 feet wide, 40 feet high; the great windowed dome, encircled by a roof garden, reached a height of 96 feet.

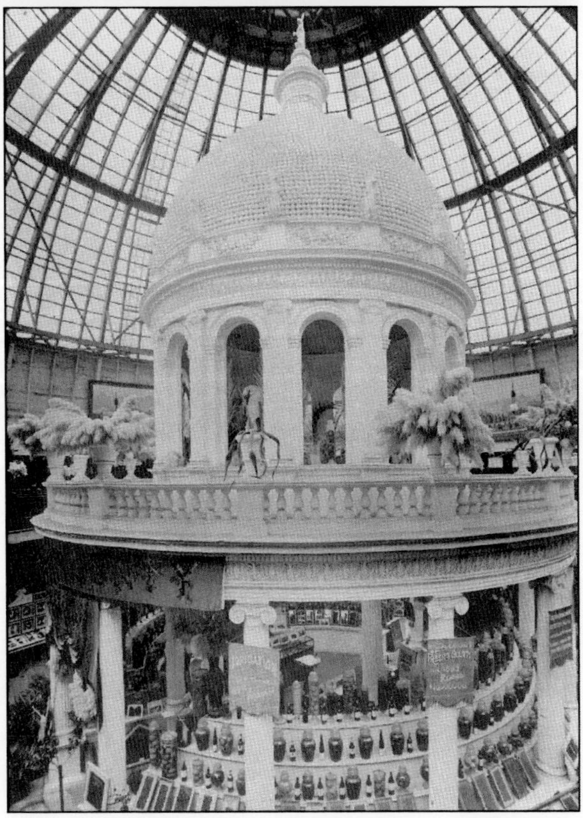

The Great Fresno Dome stood directly under the glass dome of the Agricultural Building. The lower level displayed the county's diverse produce. The upper level, modeled after the Temple of Erechtheus, contained an encircling walkway 6 feet wide, from which visitors viewed a selection of shrubs and plants. In 1894 Fresno County growers produced 103 million pounds of raisins, which caused the price to fall to 2¢ per pound.

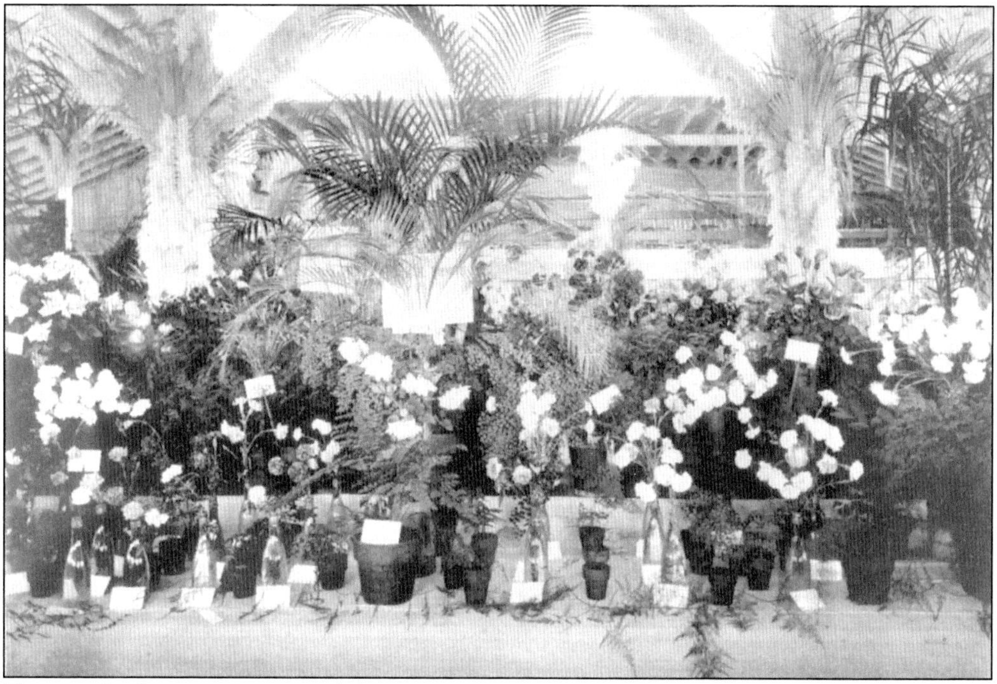

Flowers, shrubs, trees, grains and their produce: all were on display in the building, continually replenished when necessary. They usually were displayed in an exhibitor's booth, but occasionally they were grouped together by type or kind for one of the numerous shows and judgings that took place during the Exposition.

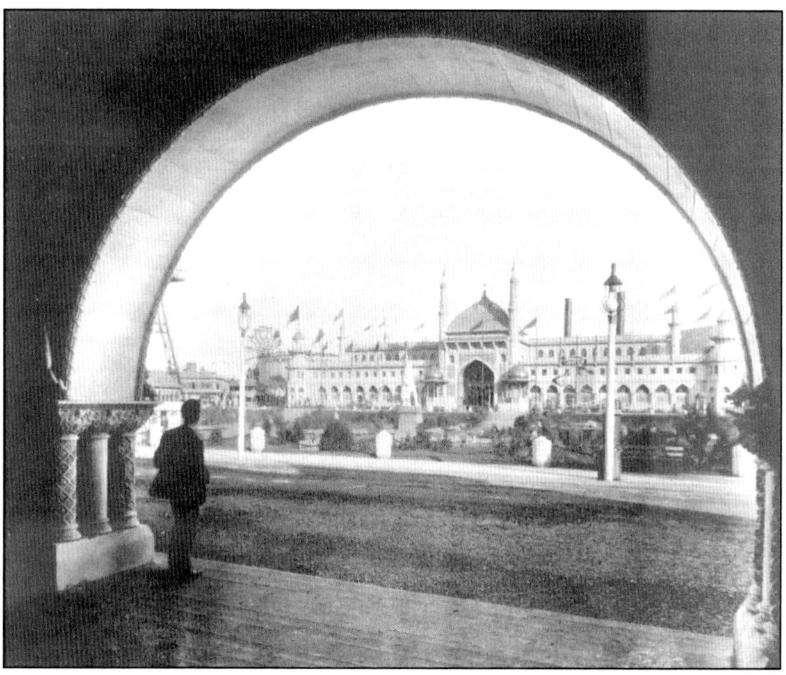

Leaving the Agriculture Building, visitors had an impressive view of the Mechanical Arts Building across the Court of Honor and the Midway beyond it. One leg of the Electric Tower may be seen to the left of the man enjoying the scene on a long ago winter day.

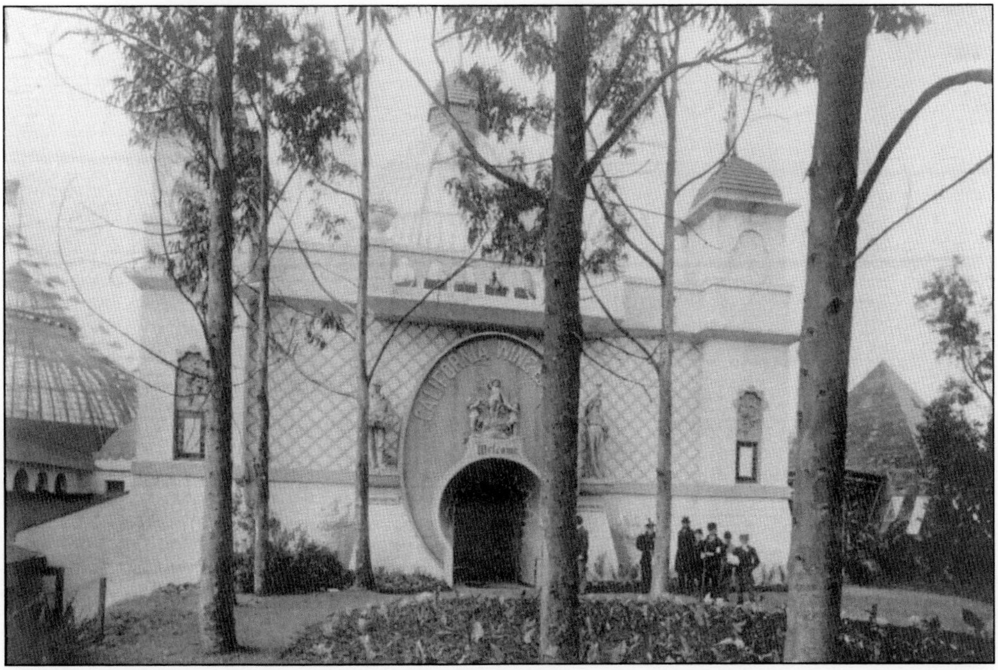

The Viticultural Building, devoted solely to California's wine industry, stood appropriately behind the Agricultural Building. Visitors entered through a doorway shaped like huge wine cask. To their left they saw a statue of a man harvesting grapes; to their right the sculpture was a woman sampling the result. The bas-reliefs showed the faces of satisfied patrons. Inside visitors saw and sampled a wide variety of wines and brandies.

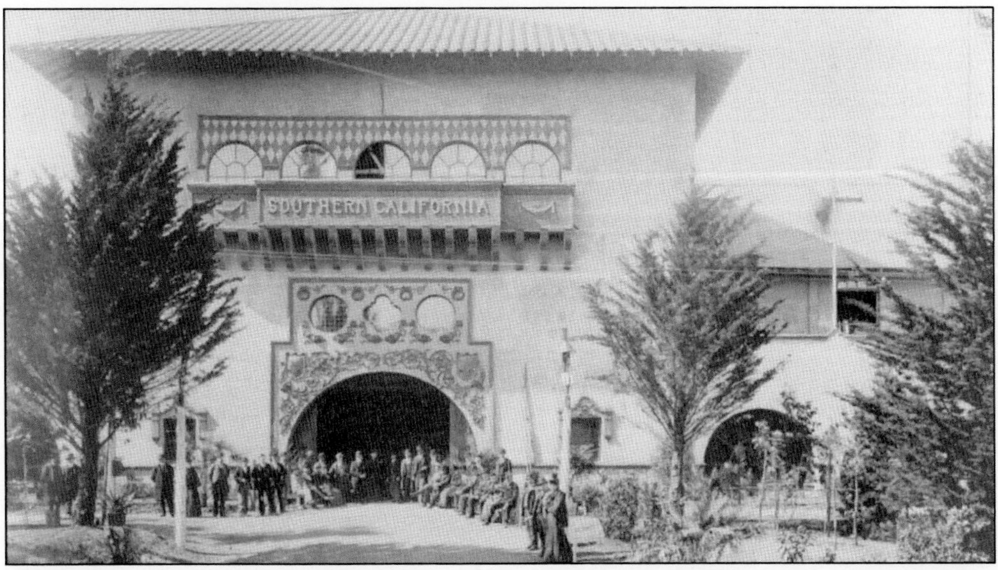

The county, state, and foreign buildings were interspersed among the buildings of the Midway. Shown here, the Southern California Building was "a happy combination of the French and Spanish, with a touch of the Moorish style so noticeable in all of the old Mission buildings." The intent was to create a style "Southern in tone and effect," underscored by the palm, orange, lemon, fig, and lime trees planted nearby.

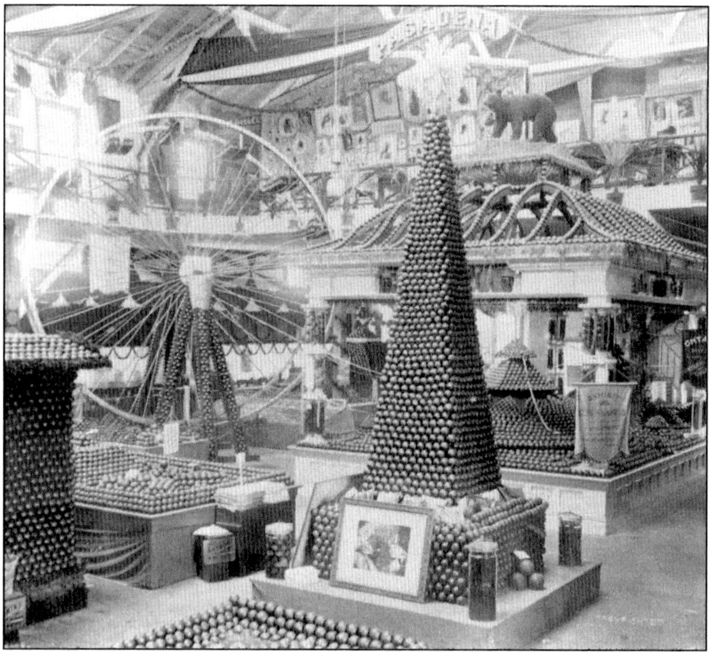

Southern California's exhibits showcased the region's agricultural abundance. Ventura County displayed its celebrated bean pagoda. Riverside County included a miniature Firth Wheel; it's 13 cars laden with oranges (*left*). Pasadena's citrus column stood nearby (*center*). Ontario's citrus gazebo had a non-citrus bear on its roof, but the roof itself was decorated with fruit (*right*). Others exhibited their produce and products in pyramids, columns, obelisks, and cones.

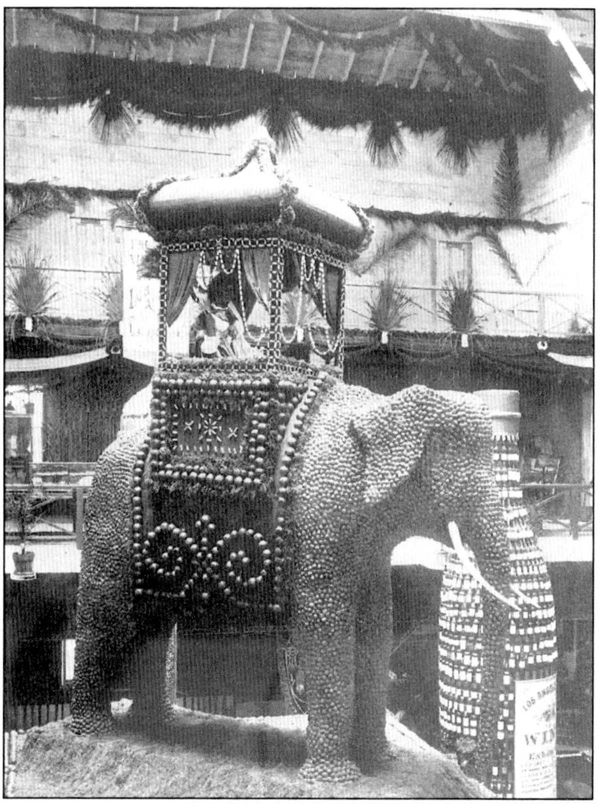

Los Angeles County's walnut elephant previously had been seen at the Columbian Exposition. Standing on a triumphal arch of oranges, only the elephant itself consisted of walnuts—1,000 pounds of them. The saddlecloth, howdah, and trappings were outlined in citrus fruits, peanuts, and corn. The contented passenger was made of wax. The County claimed to raise more grapes, oranges, almonds, olives, and walnuts than anywhere else in the world.

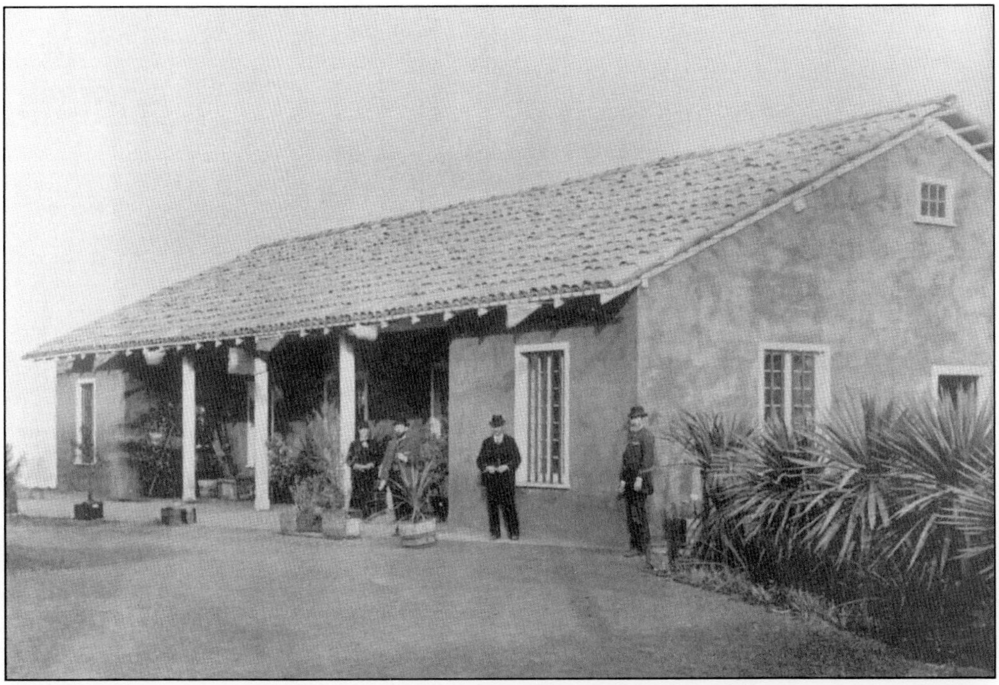

The Monterey County Building looked like a Spanish California ranch house, common 50 years earlier, but rare by the time of the Exposition. Outside visitors saw the flagpole used by John C. Fremont to hoist the stars and stripes for the first time in California. Nearby was the old cannon that fired the first salutes. Inside they could browse a collection of handiwork produced for the friars by Native Americans.

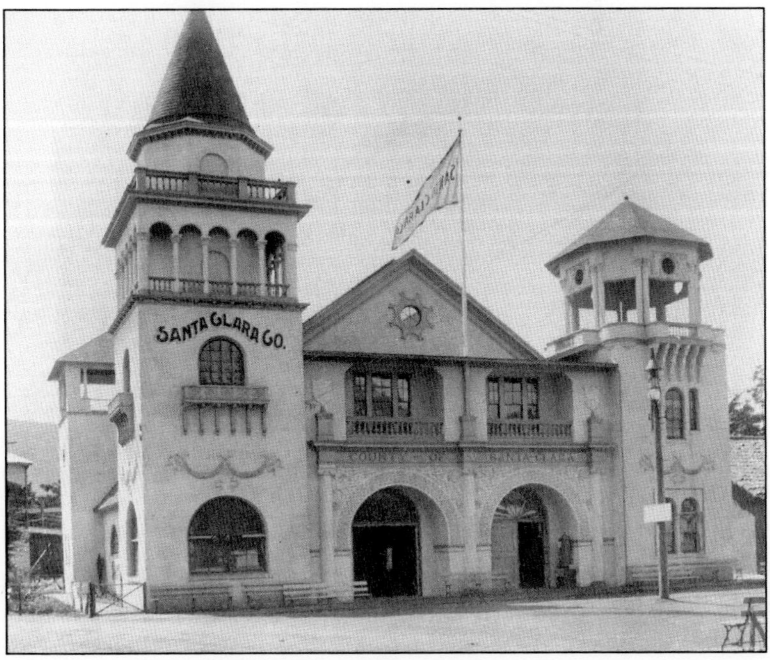

The Santa Clara County Building housed a display that dazzled visitors with the agricultural wealth of the region, then probably the richest in the state. Famous for its wheat, wine, and orchards, 90 percent of all the prunes produced in California came from its groves. Silicone chips and semi-conductors have replaced the grapes, plums, and apricots.

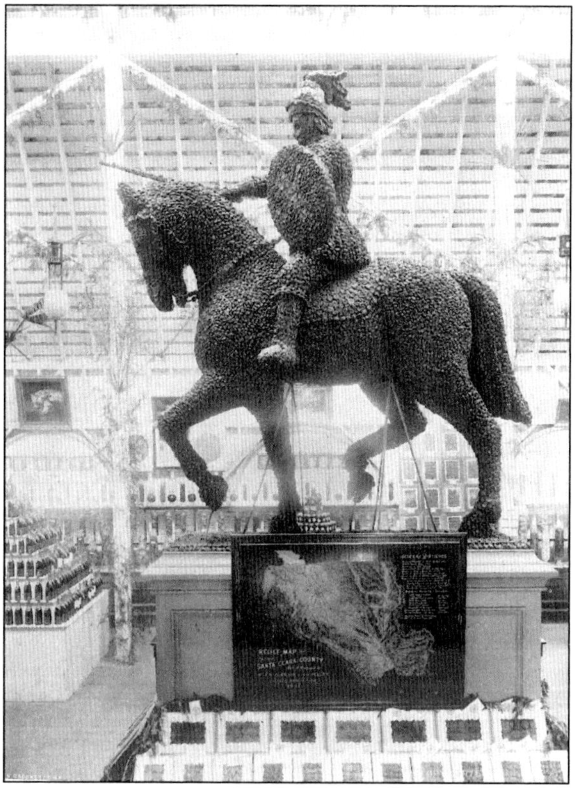

The display inside the Santa Clara County Building featured the statue of a charging knight whose figure, shield, helmet, plume, sword, horse, and bridle were made almost entirely of prunes—a few apricots, plums, and nectarines were used for highlights. He proudly pointed to a selection of local wines, whose producers stated, here and elsewhere, "Only second to her mines are California's wines."

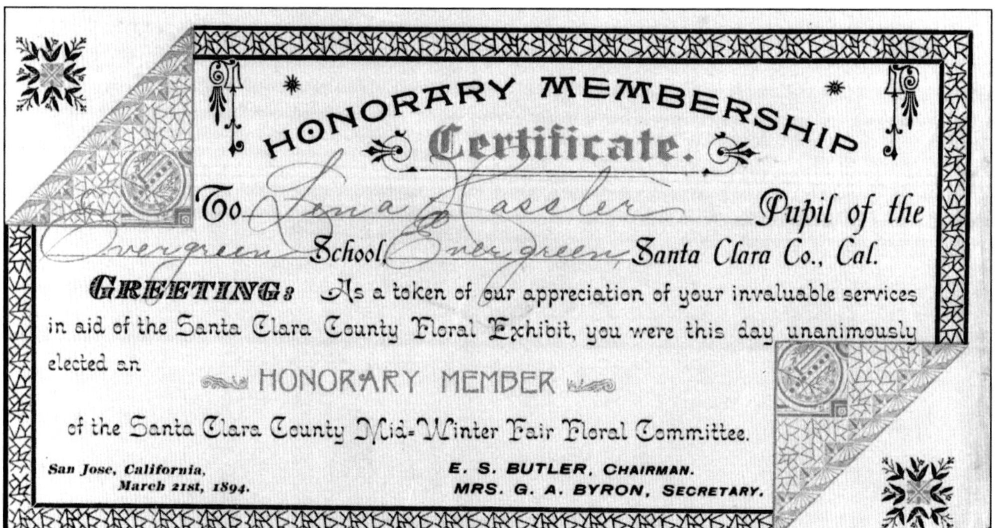

HONORARY MEMBERSHIP

Certificate.

To Lena Hassler Pupil of the Evergreen School, Evergreen, Santa Clara Co., Cal.

GREETING: As a token of our appreciation of your invaluable services in aid of the Santa Clara County Floral Exhibit, you were this day unanimously elected an

HONORARY MEMBER

of the Santa Clara County Mid-Winter Fair Floral Committee.

San Jose, California,
March 21st, 1894.

E. S. BUTLER, CHAIRMAN.
MRS. G. A. BYRON, SECRETARY.

The Fair recognized many people and communities for their contributions to its success. Lena Hassler of Evergreen School, Santa Clara County received a certificate of appreciation for her "invaluable services in the aid of the Santa Clara County Floral Exhibit," which made her an honorary member of the County's Midwinter Fair Floral Committee. Given all the spring cleanings it survived before becoming a collector's item, she surely treasured the citation.

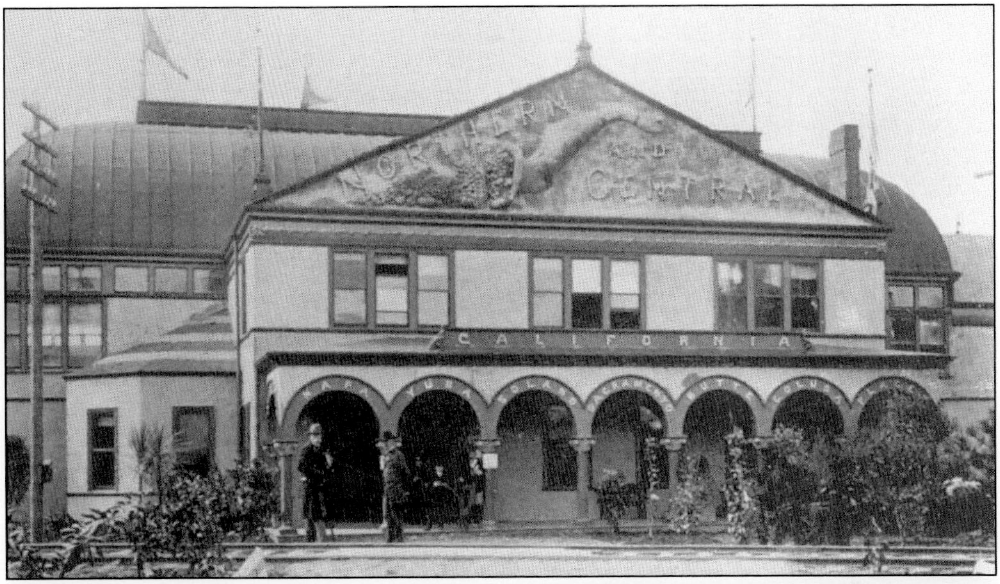

The Northern and Central California Building contained displays from a dozen counties, their names modeled onto the arches of the porch. The horn of plenty on the building's pediment announced the theme of the exhibits inside: the region's great natural wealth. Arrays of woods and minerals, deciduous fruits, hops, wine, wool, and citrus, among others, showed the diversity of product.

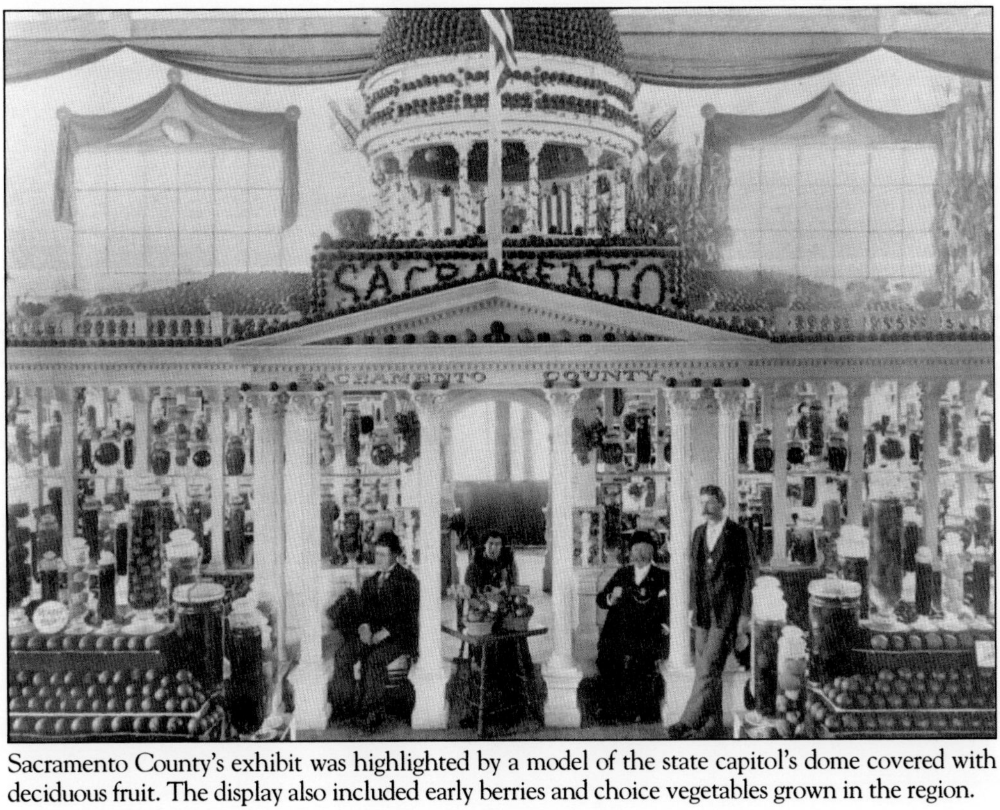

Sacramento County's exhibit was highlighted by a model of the state capitol's dome covered with deciduous fruit. The display also included early berries and choice vegetables grown in the region.

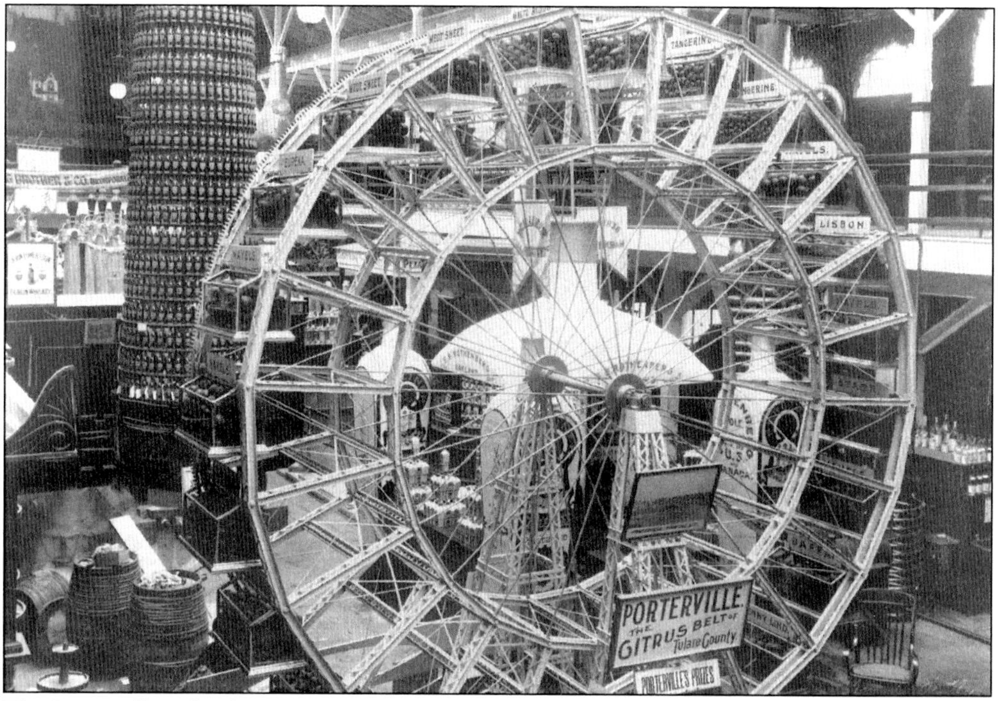

The Porterville Wheel, part of Tulare County's exhibit, was another working model of the Firth Wheel. Here the passengers were raisins, oranges, lemons, prunes, figs, walnuts, peanuts, and almonds. The district won several first premium awards at the Fair, including best exhibit of citrus fruit and best Villa Franca lemons.

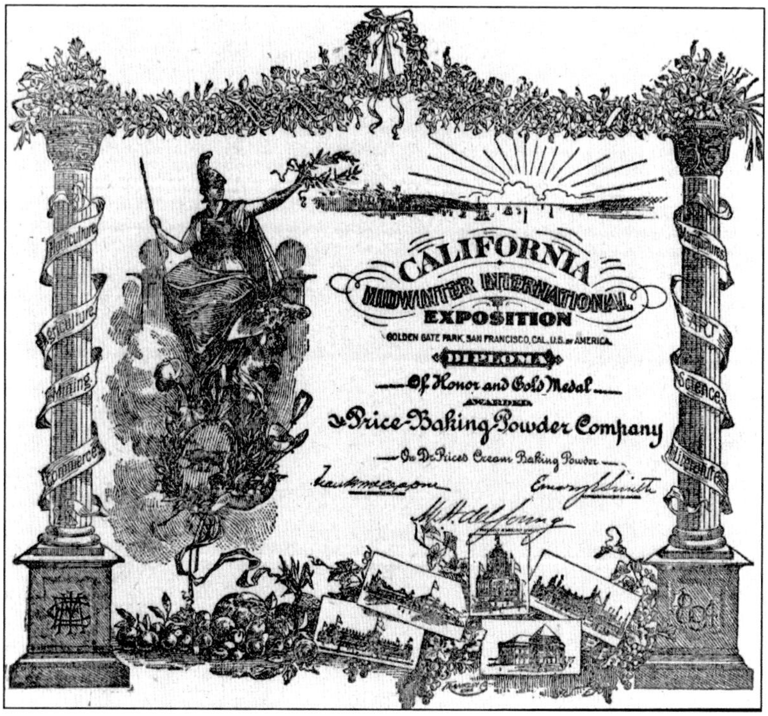

The Fair honored 2,083 exhibits for excellence. Winners received an award ribbon, a bronze medal, and a diploma of honor. Exhibit judges, commissioners, and other dignitaries also took home diplomas. At the time, companies valued such recognition highly; many put information about it on business cards, correspondence, and product labels. Some have it there still.

Each prize-winning exhibit received a bronze medal in a silk-lined case. Had they won a second or first place citation—and were willing to pay the cost—exhibitors could order a silver, gold-filled, or 14k gold medal. Few did. Typical of the times, the design by G. Giaccomi used as much symbolism as possible. It is the only three-part award medal ever issued by an American Exposition.

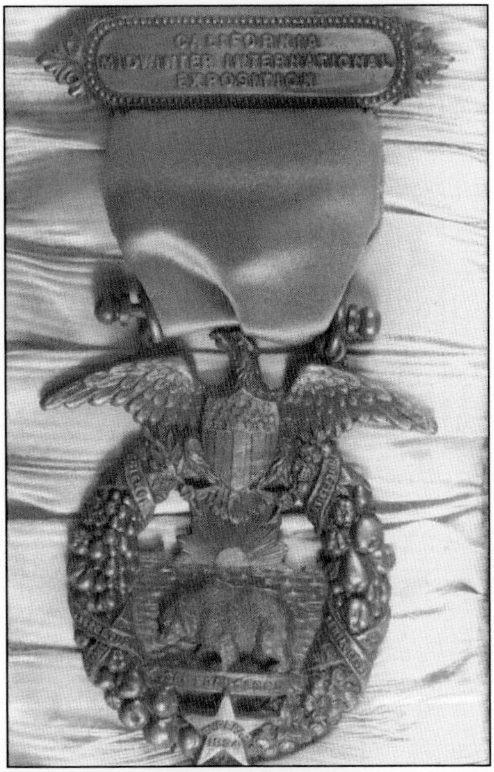

Visitors to the San Mateo County Building saw exhibits of agriculture and horticulture. Especially impressive was the display of wood products, the county being home to 100,000 acres of redwoods, almost a third of its total area. Its output of lumber, shingles, and other items made of wood was worth $1 million a year.

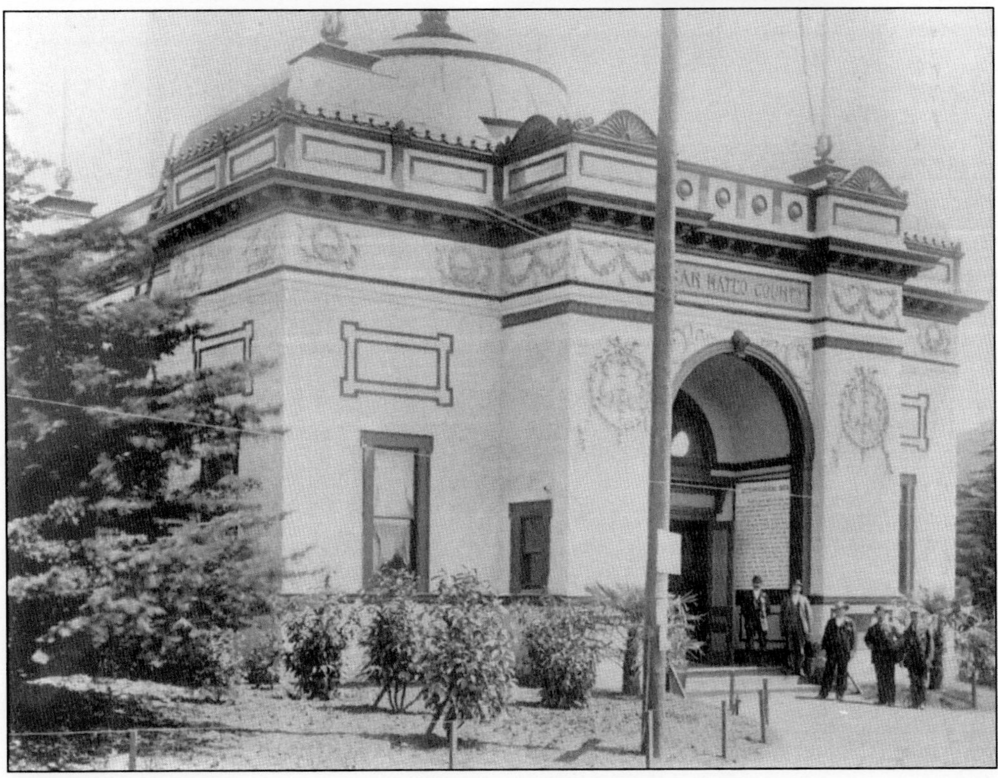

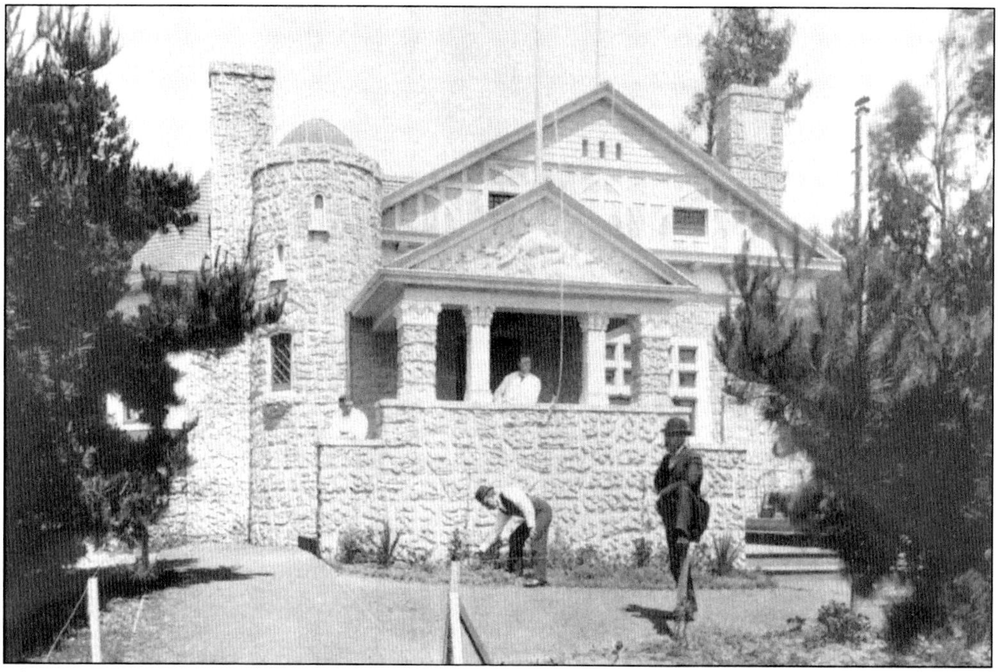

The Canada Club Building was the social headquarters of resident and visiting Canadians at the Fair. Chicago's extensive Canadian exhibit went to the Antwerp Exposition of 1894, not to San Francisco, but local organizers secured a mineral display for the Mechanical Arts Building and a collection of manufactured goods for the Manufactures Building. Canadian Day was Friday, April 2.

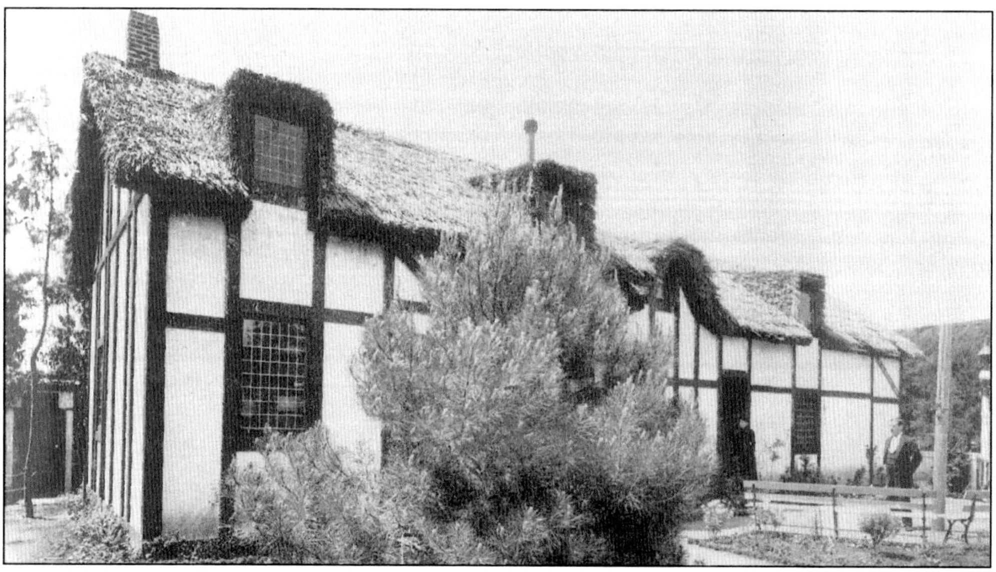

Ye Ann Hathaway Cottage, erected by British subjects living in San Francisco, functioned as a headquarters for them, British visitors to the Exposition, and guests. It contained no exhibits. A faithful copy of Shakespeare's Stratford-Upon-Avon home, both outside and within, it was designed to create feelings of restful harmony and quiet dignity that, according to Elizabethan gossip, were seldom found in the original.

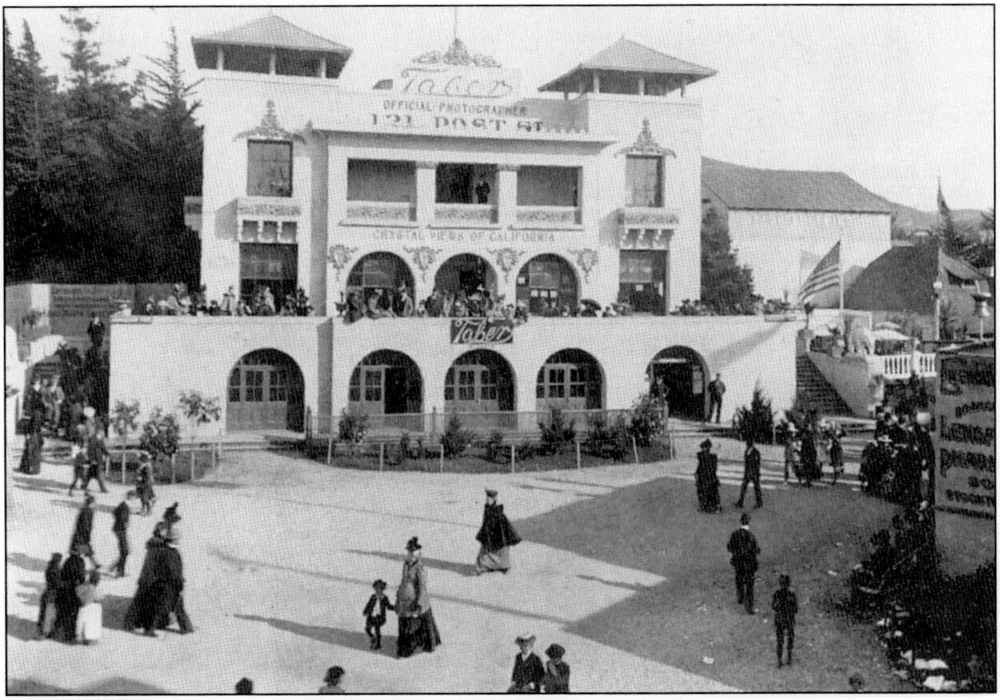

Isaiah Taber had exclusive rights to photograph the Fair, although amateur photographers could take pictures for their personal use. His studio on the Exposition grounds sold stereoviews and photographs of famous people and places. The Great Earthquake of 1906 destroyed his entire inventory of glass plates, view negatives, and portraits on glass, ending his career. A street in the once photographer's district, south of Market, is named for him.

There was much more to the Court of Honor than educational exhibits. Parades, concerts, and daredevil performances took place regularly there. Here, in an amateur photograph taken from a glass slide, an aerialist performs his high wire act for onlookers. The trickiest part was maneuvering down the spiral ramp.

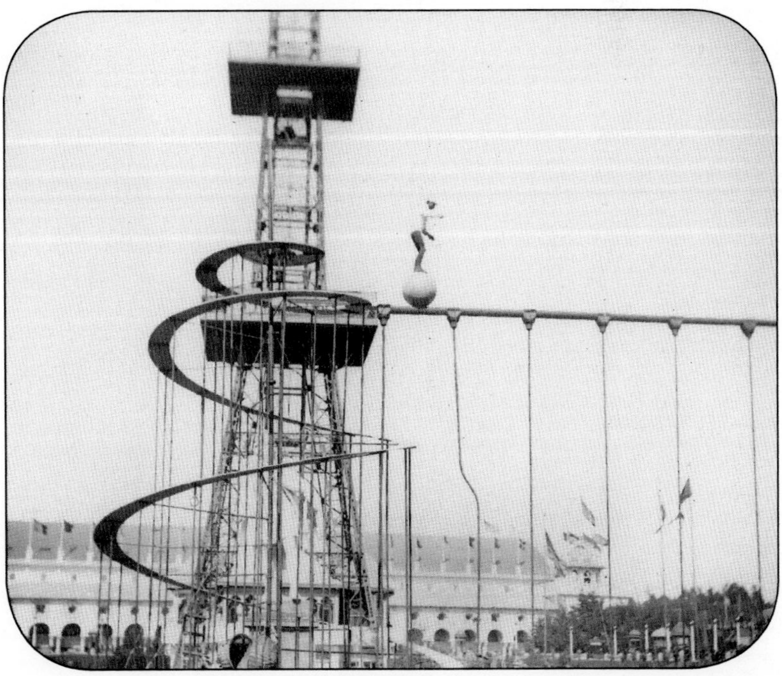

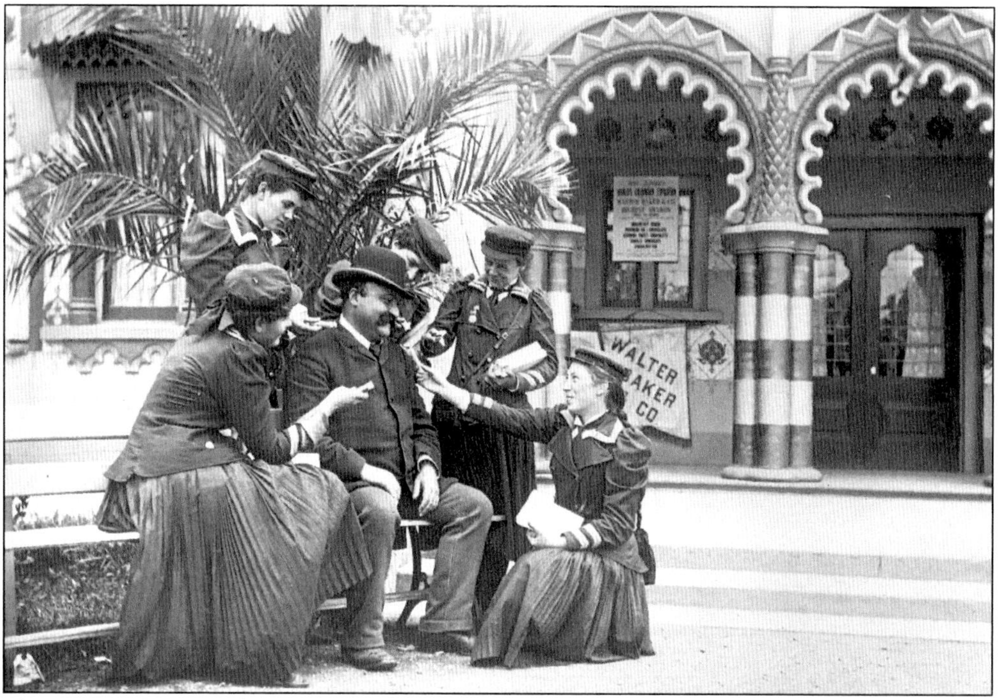

The Court of Honor also was home to the Gum Girls, who strolled the fairgrounds, enticing fairgoers to purchase their wares. Who could refuse their plaintive call, "Gum? Buy gum?" Asked if she liked her job, one girl replied, "What der yer s'ppose? That's the way I make a livin'—by gum!" So much public attention seems to have delighted, embarrassed, or overwhelmed our very buttoned up Victorian gentleman.

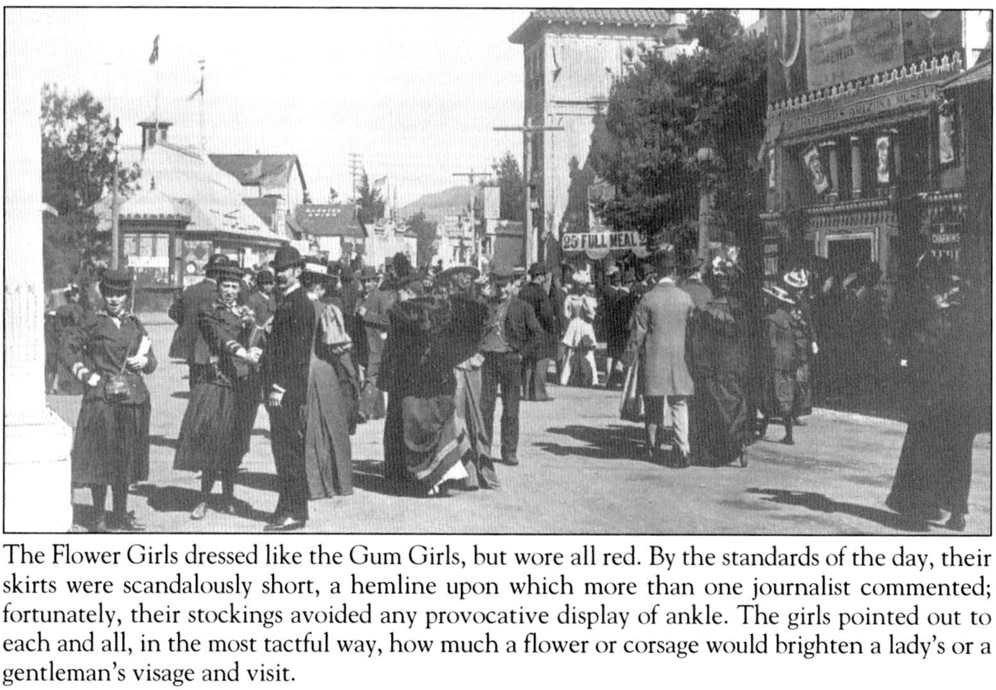

The Flower Girls dressed like the Gum Girls, but wore all red. By the standards of the day, their skirts were scandalously short, a hemline upon which more than one journalist commented; fortunately, their stockings avoided any provocative display of ankle. The girls pointed out to each and all, in the most tactful way, how much a flower or corsage would brighten a lady's or a gentleman's visage and visit.

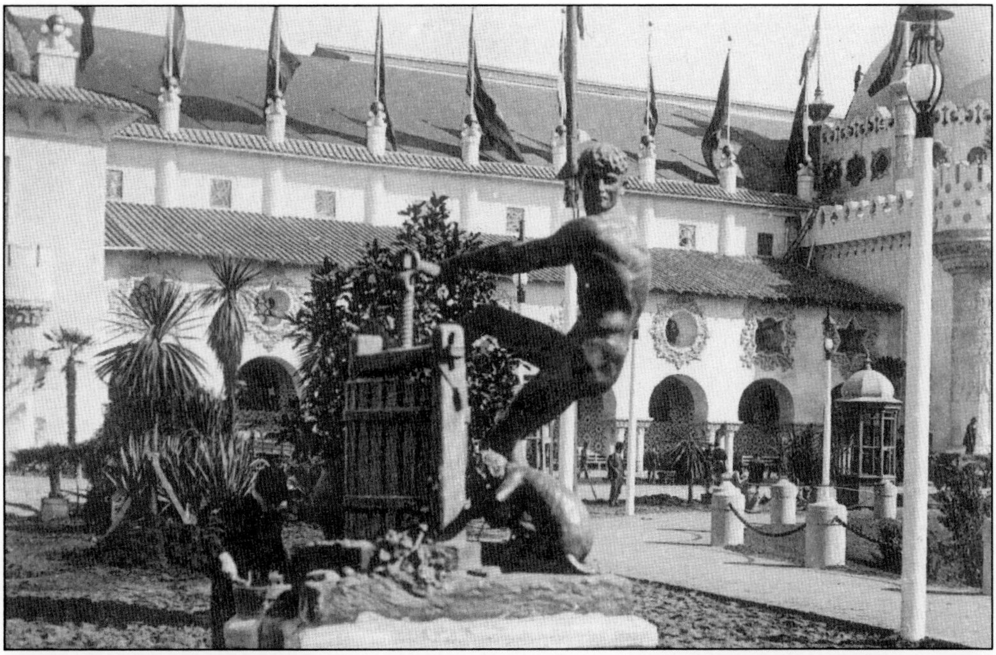

The Cider Press by Thomas Shields Clarke won a gold medal at the Fair. Actually it provided water, not cider, for thirsty fairgoers; how many were disappointed to discover this wasn't documented. Although the bucket went missing long ago, the master of the press and his helper are still in Golden Gate Park, for more than 100 years offering their gift of refreshment to parched passersby.

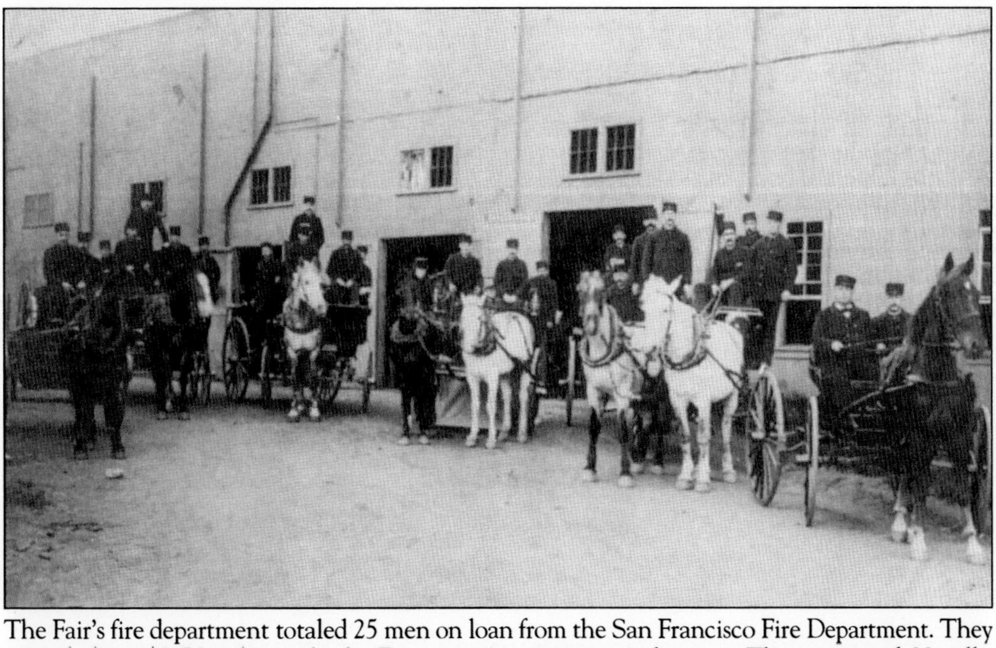

The Fair's fire department totaled 25 men on loan from the San Francisco Fire Department. They earned about $2.56 a day; only the Exposition's managers made more. They answered 22 calls, mostly false alarms. The greatest loss of property occurred on June 18, when two cottages in the Dahomey Village burned to the ground. The loss totaled $150. Fortunately, no one was injured.

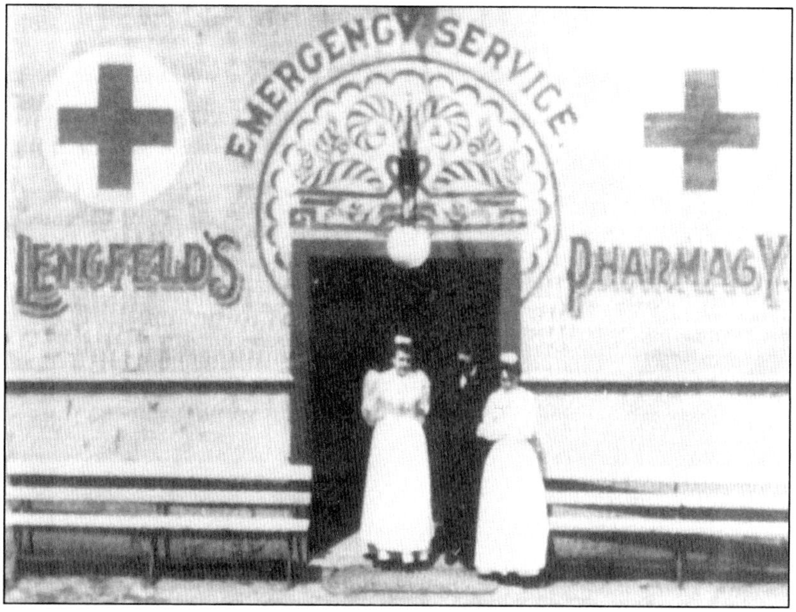

The Exposition's hospital was modern, clean, and efficient. Martin Regensberger, the resident physician, couldn't find an ambulance locally, so de Young obtained the one built by the Studebaker Wagon Company for the Chicago Fair. When the Exposition closed, Theresa Fair, future owner of San Francisco's Fairmont Hotel, bought the ambulance and donated it to the city, creating the first and oldest continuously operating municipal ambulance service in the U.S.

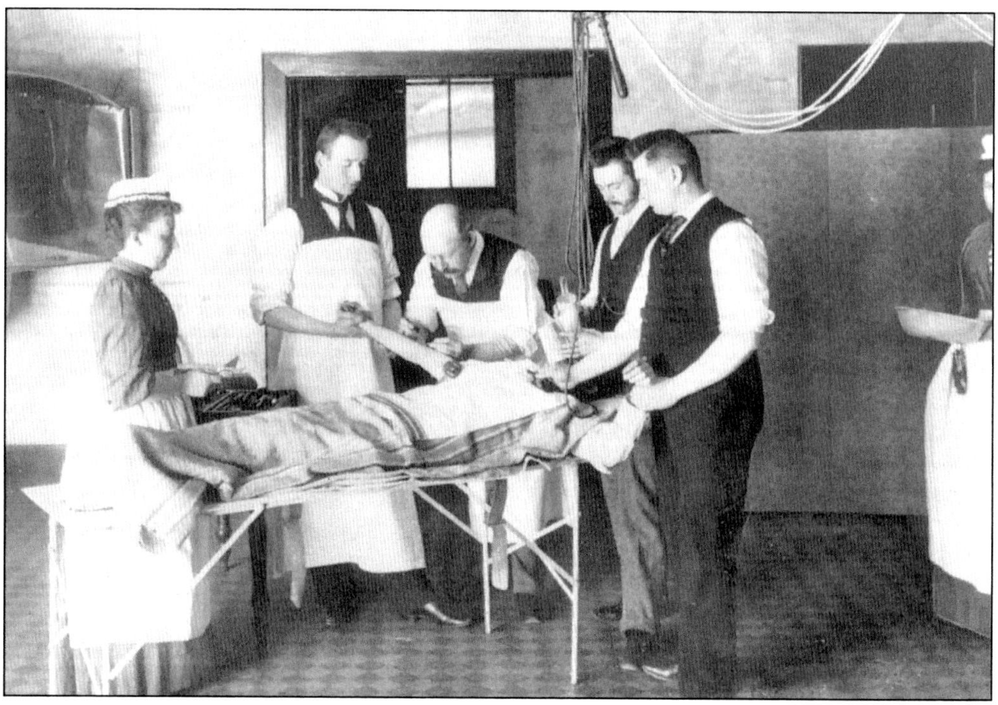

The hospital's small staff treated 1,935 cases. Most were minor, primarily traumatisms, abscesses, and migraines. Fairgoers also suffered 16 fractures, 2 concussions, 38 burns, 53 sprains, and 4 dislocations. For healthy visitors the facility became its own exhibit of a modern clinic, complete with a first aid station, an operating room, and a pharmacy.

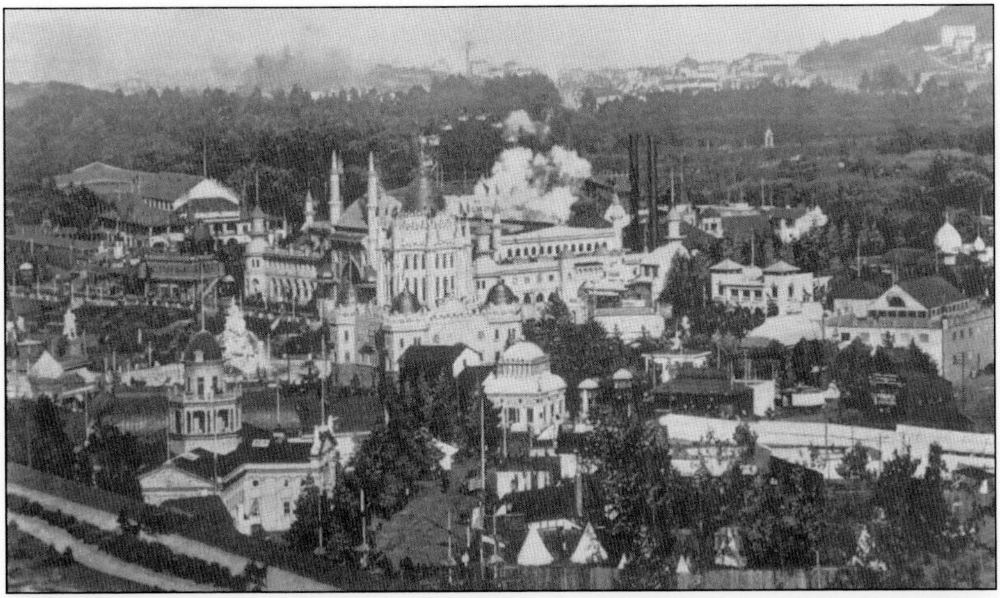

From a distance, one building close upon another, the Fair appears very much the jumble of architecture that its critics disdained. The Administration Building, here just left of center, seems part of the Mechanical Arts Building behind it. Regardless, it looks like a wonderland to explore.

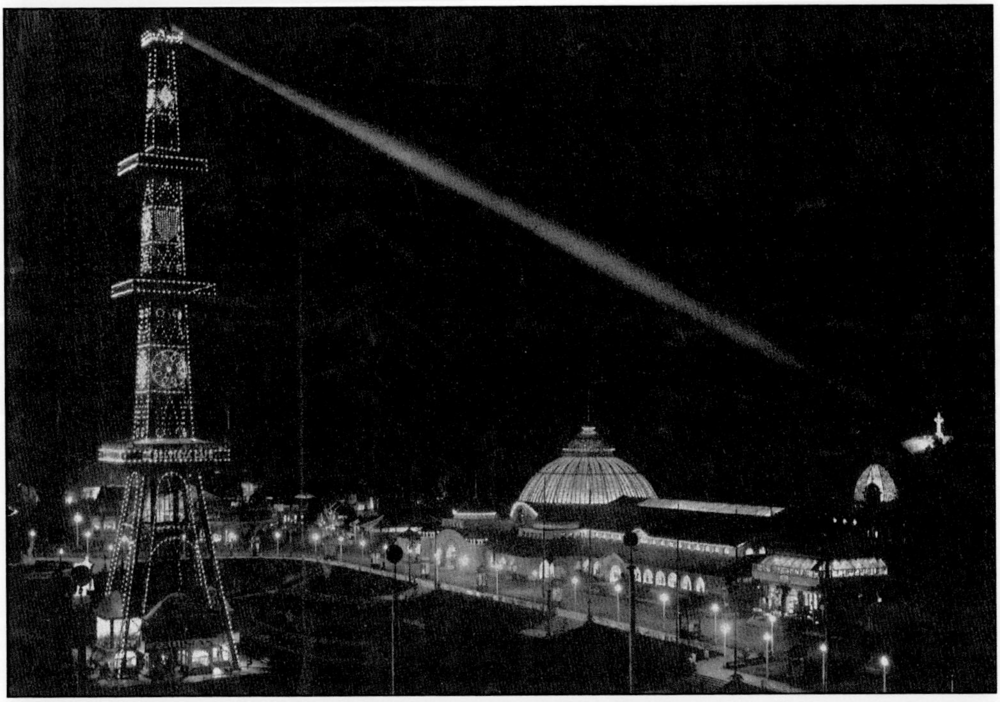

The fairgrounds at night took on the look of an entirely different place. Most homes were still lit with kerosene or gas. Electric lighting wasn't a novelty to San Franciscans, even though the city still used 4,300 gas street lamps as late as 1927. It still was an amazement to many visitors from small towns and rural communities; some had never before seen a working incandescent light bulb.

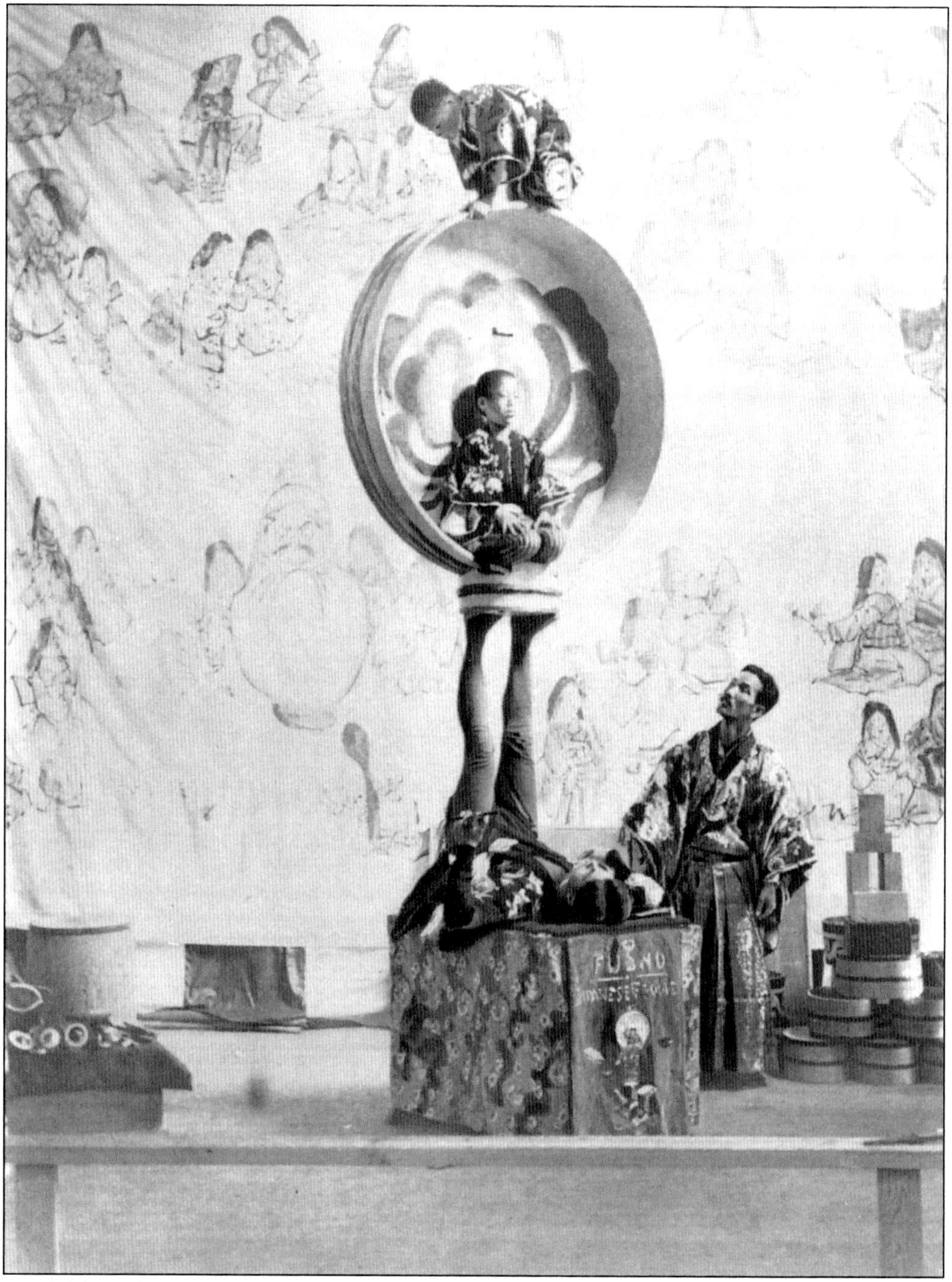

Youthful acrobats perform in the theater of the Japanese Village. Their consummate skills made their show one of the most popular at the Exposition.

Three

A GATHERING OF PEOPLES

Some of the most popular attractions at the Midwinter Fair were the exhibits of people. Similar presentations had been well received at the Columbian Exposition and would continue to be featured at American expositions for years to come. Native Americans, Japanese, Egyptians, Aleutians, Hawaiians, Turks, South Seas Islanders, and Chinese—all had their villages along the Midway, neighbors of the St. Bernard dog team, the Mystic Mirror Maze, and the Animal Arena.

Some fairgoers found all this enlightening. "Thousands of sightseers at the Midwinter Fair can mingle their 'Oh, mys!' of admiration before a world-famous painting, or an electric tower, with a merry ripple of laughter at the recollection of the antics of the little people on the Midway," wrote the *Overland Monthly*. "The Midway is an educator in its way. It affords peeps into lands and customs that have hitherto been myths as far away and impalpable as the man-in-the-moon."

Other visitors bemoaned the corruption of distant peoples by American culture. "We demand that our foreigners shall be simon pure" wrote one critic, perhaps tongue in cheek, "The combination of the Japanese kimono, American shoes, and a derby hat is not pleasing."

At the same time that the Fair's patrons gawked or marveled at people of other cultures, finding them savage or admirable, ridiculous or impressive, threatening or noble—or simply different—their leaders were implementing policies to keep them either out of mainstream American society or out of the U.S. entirely.

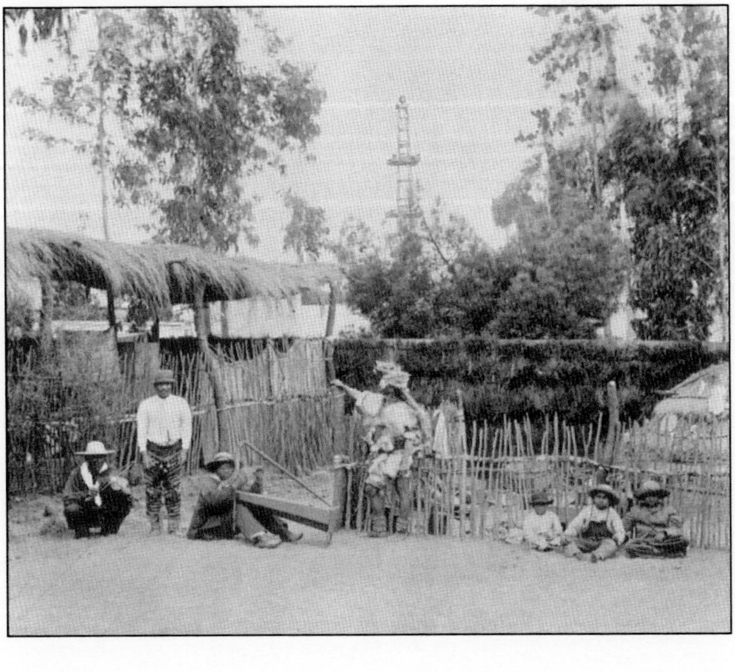

The Arizona Indian Village was "a bit of desert life transported from the rude frontier to the very center of the world's civilization," where its residents found the winter weather a bit brisk for their tastes. The inhabitants of this miniature municipality lived in wickiups constructed of grass when not performing the Deer Dance for visitors, who also saw a collection of woven baskets and "a forest of giant cacti."

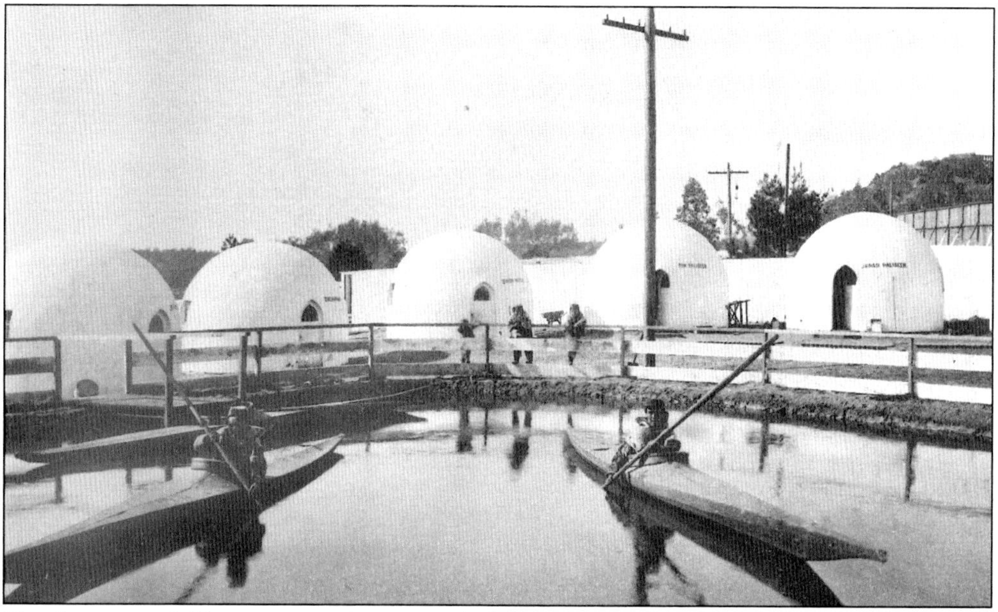

The Esquimaux Village presented an entirely different Native American community, this one complete with dog teams, papier-mâché igloos, and a miniature lake. Required to wear hoods, parkas, boots, and other warm clothing, the residents naturally thought the local climate was much too toasty. Still, it was better than their summer in Chicago, where most of the group tossed aside both furs and jobs because of the Midwestern heat.

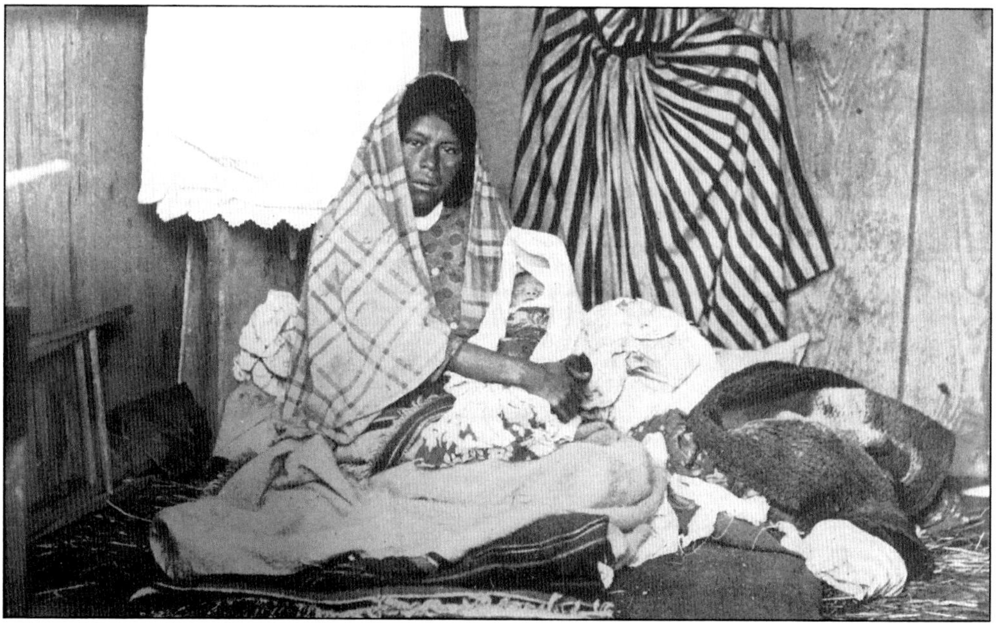

An unplanned attraction came about when an Aleutian gave birth. Here four-day-old Francisca Examiner Deer poses with her mother, made possible because "a peculiarity among uncivilized people is the fact that maternity is accompanied by neither pain nor weakness." Surely the mother was pleased to welcome visitors, "appreciative as she was of the acute desire upon the part of white people to see infants of uncivilized mothers."

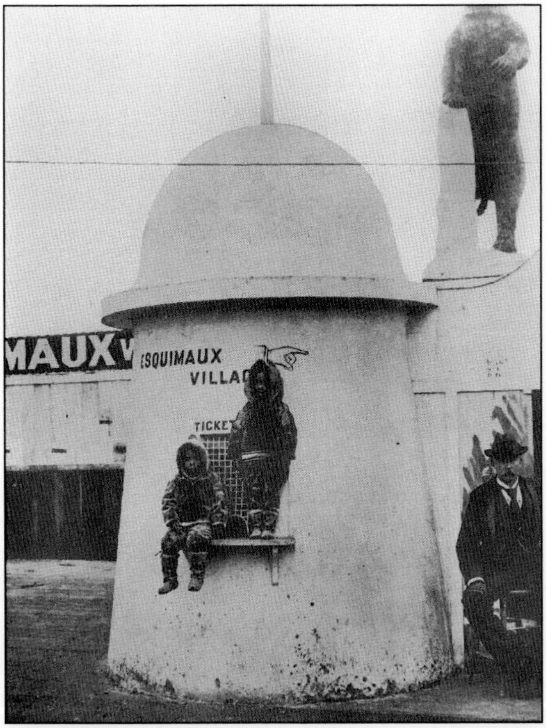

Two of the older children of the village pose for the photographer at the ticket booth. Dressed in their parkas, waterproof pants and boots for a winter in the extreme northern latitudes, which fairgoers found immensely interesting, they look completely up to date for the trip to school on a snowy day in any modern American city.

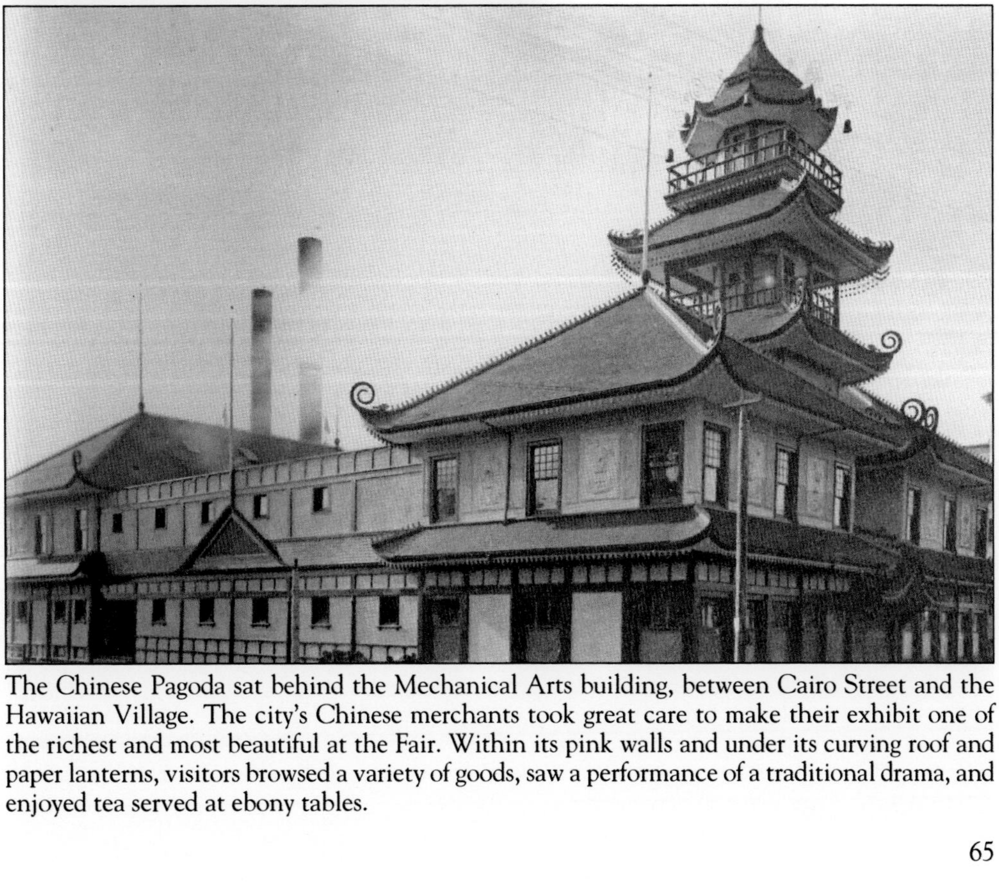

The Chinese Pagoda sat behind the Mechanical Arts building, between Cairo Street and the Hawaiian Village. The city's Chinese merchants took great care to make their exhibit one of the richest and most beautiful at the Fair. Within its pink walls and under its curving roof and paper lanterns, visitors browsed a variety of goods, saw a performance of a traditional drama, and enjoyed tea served at ebony tables.

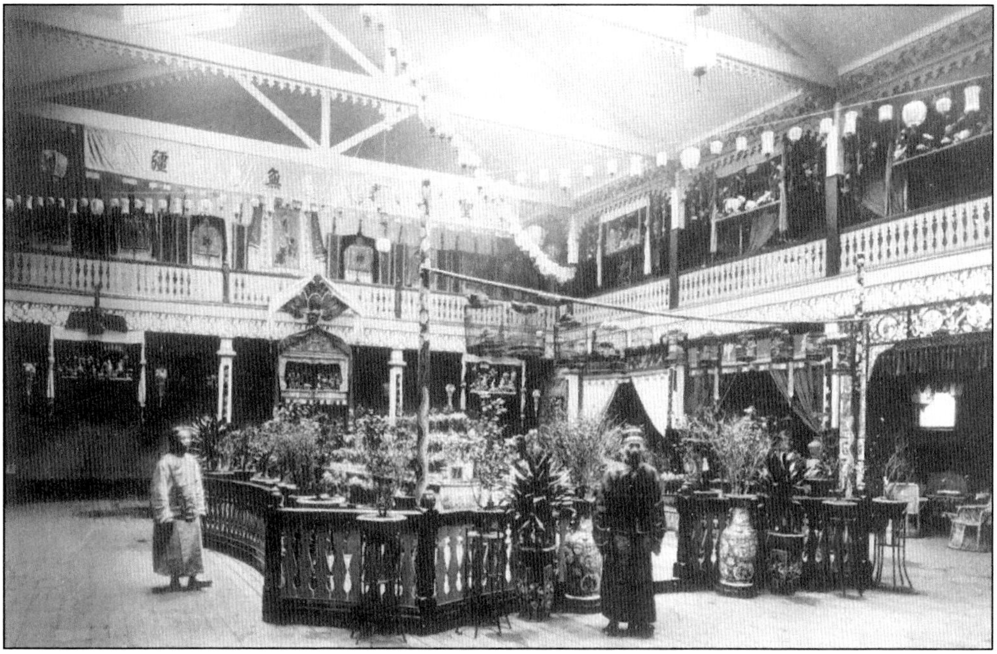

The display of China's products included tableware, silk, ivory carvings, filigree, silver, and coral items. "There's nothing finer at the Fair / Than their imported chinaware," wrote versifier Daniel Oscar Loy in his *Poem of the Golden State and Midwinter Exposition*. Later in the year, Sun Yat-Sen, living in Honolulu, established the Xingzhonghui to promote educational and agricultural modernization in his homeland.

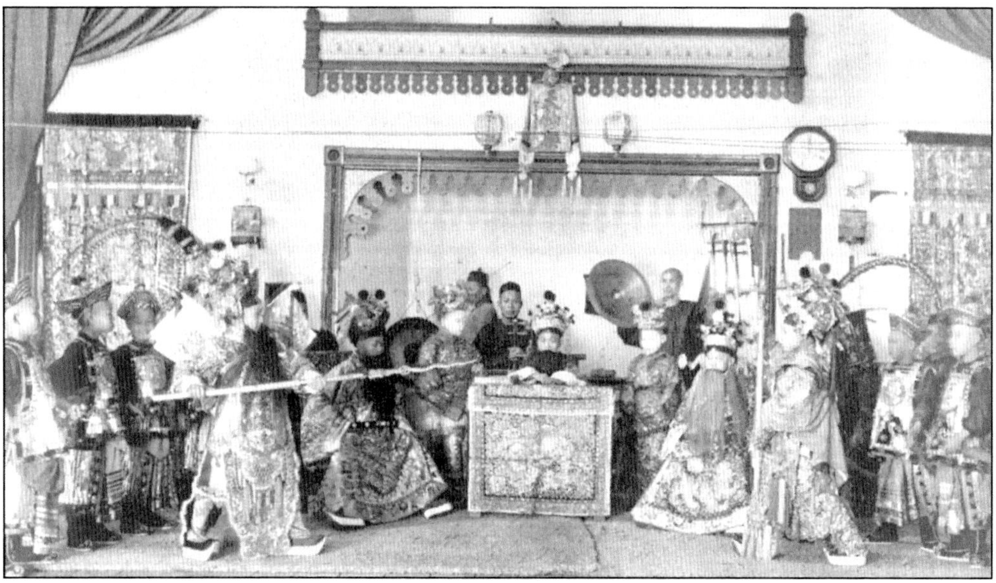

The Chinese Theater featured a troupe of 12 adults and 20 children. They presented a repertoire of plays with themes from either the mythology or history of China. One reviewer for the San Francisco *Examiner* wrote that, although "it kills enough people to satisfy the most exacting gallery god," its "several acts drag to a conclusion" because "its dialogue is interminable and the action slow."

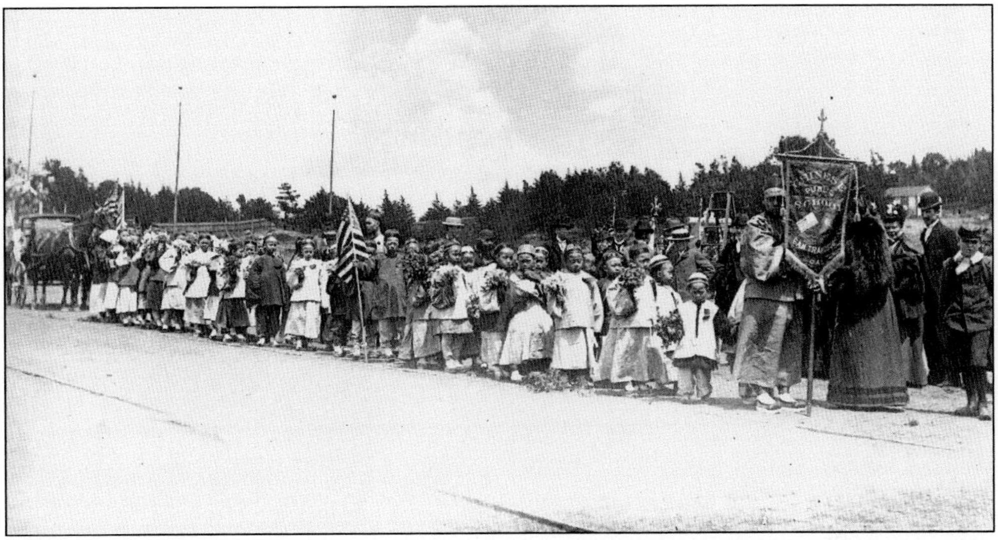

A Chinese Parade makes its way through the fairgrounds. The children, carrying American flags, are from the city's Chinese Public School. Six weeks before the Exposition closed, the U.S. and China, replacing an earlier pact, agreed to prevent Chinese workers from entering the country for 10 years; Congress extended the law twice before the Immigration Act of 1924 essentially excluded all Asians. The ban finally was repealed in 1943.

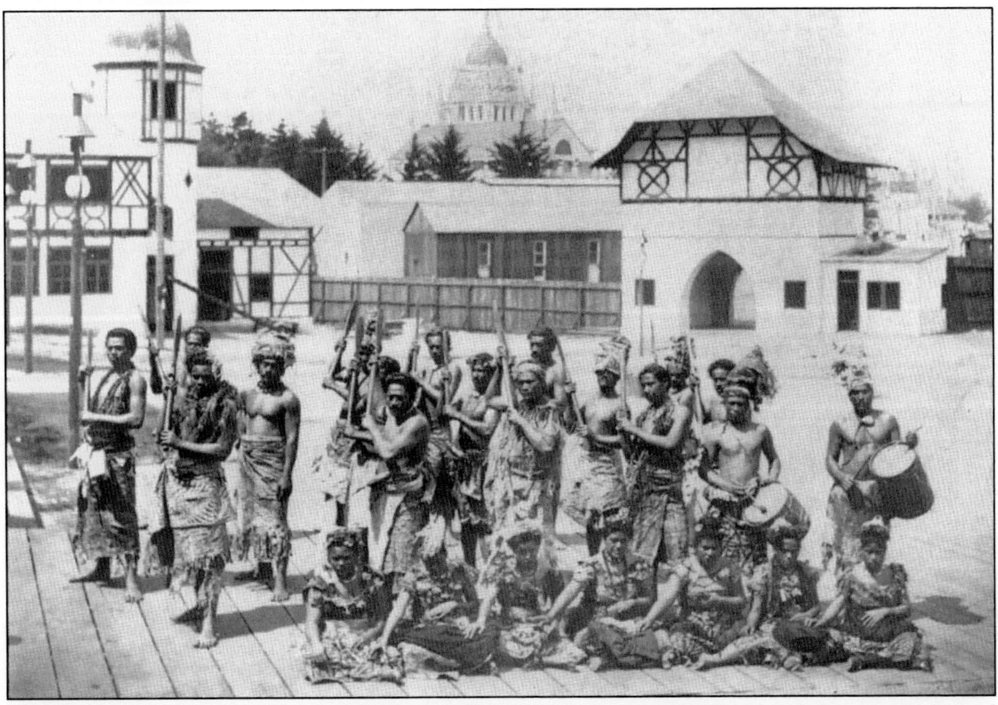

The South Seas Island Village included people from Samoa and Wallis Island. Fairgoers admired their houses made entirely of palm leaves, grasses, and bamboo. The Islanders performed traditional songs, religious ceremonies, and festive dances, which many pathway pundits found unusual. Were the Islanders cooperative? "My hardest task," stated the concessionaire, "is to make these people wear their clothes."

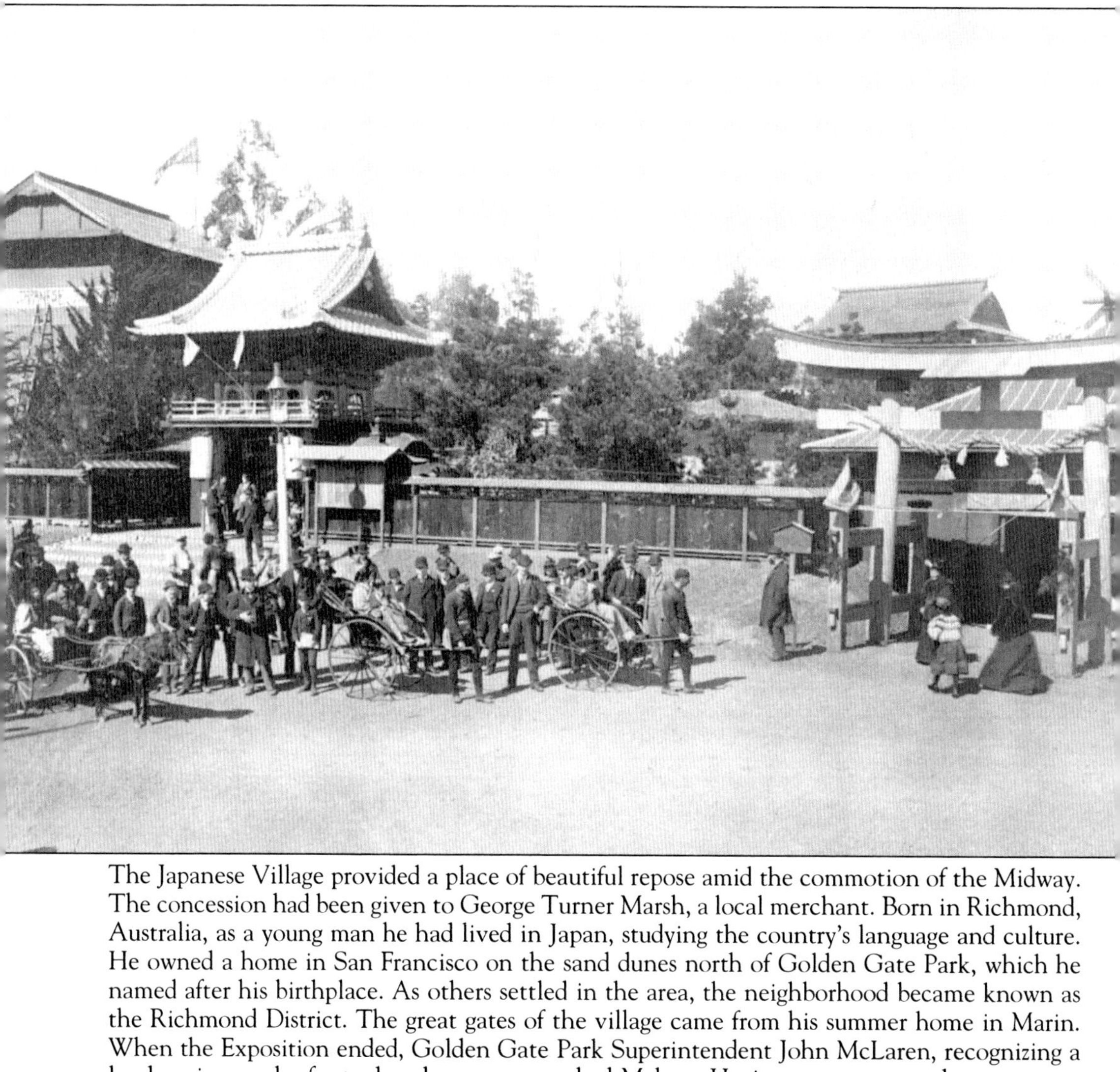

The Japanese Village provided a place of beautiful repose amid the commotion of the Midway. The concession had been given to George Turner Marsh, a local merchant. Born in Richmond, Australia, as a young man he had lived in Japan, studying the country's language and culture. He owned a home in San Francisco on the sand dunes north of Golden Gate Park, which he named after his birthplace. As others settled in the area, the neighborhood became known as the Richmond District. The great gates of the village came from his summer home in Marin. When the Exposition ended, Golden Gate Park Superintendent John McLaren, recognizing a landscaping work of art when he saw one, asked Makoto Hagiwara to convert the temporary exhibit into a permanent Japanese Tea Garden for the park. Hagiwara's family lived at the Garden, maintained it, and enhanced it until 1942, when they and other Americans of Japanese descent were forced to relocate to concentration camps. It is the oldest public Japanese garden in the U.S.

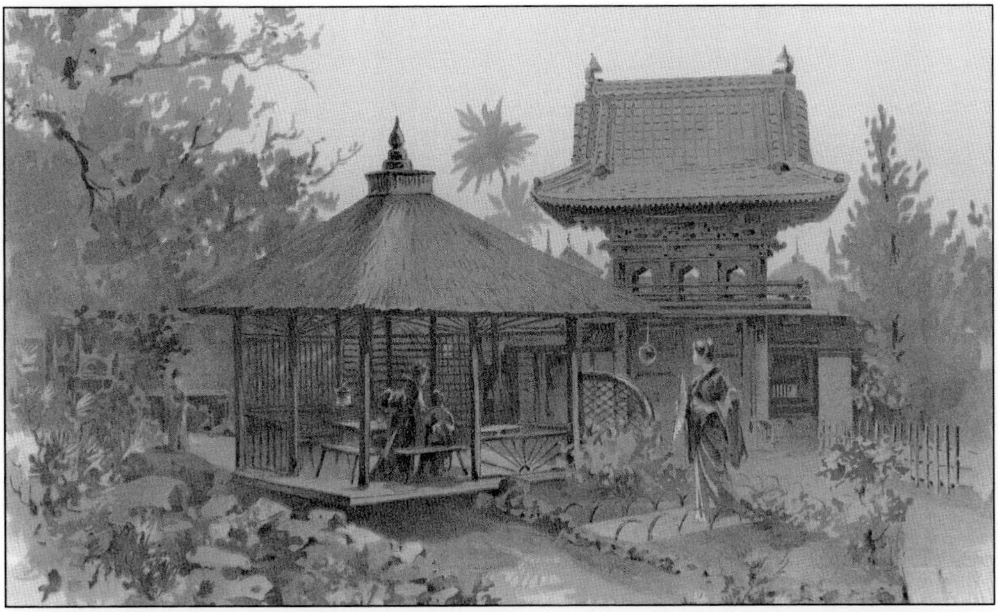

The site for the Japanese Village was chosen because of its irregular grounds and its fir trees, which also grew in Japan. Unfortunately, the Japanese fir's needles are fewer and further apart than their American counterparts, so Marsh hired workers to thin them. Such devotion to detail, combined with the gracefulness of the landscaping and the graciousness of kimono-clad residents, made the village the most popular concession of the Fair.

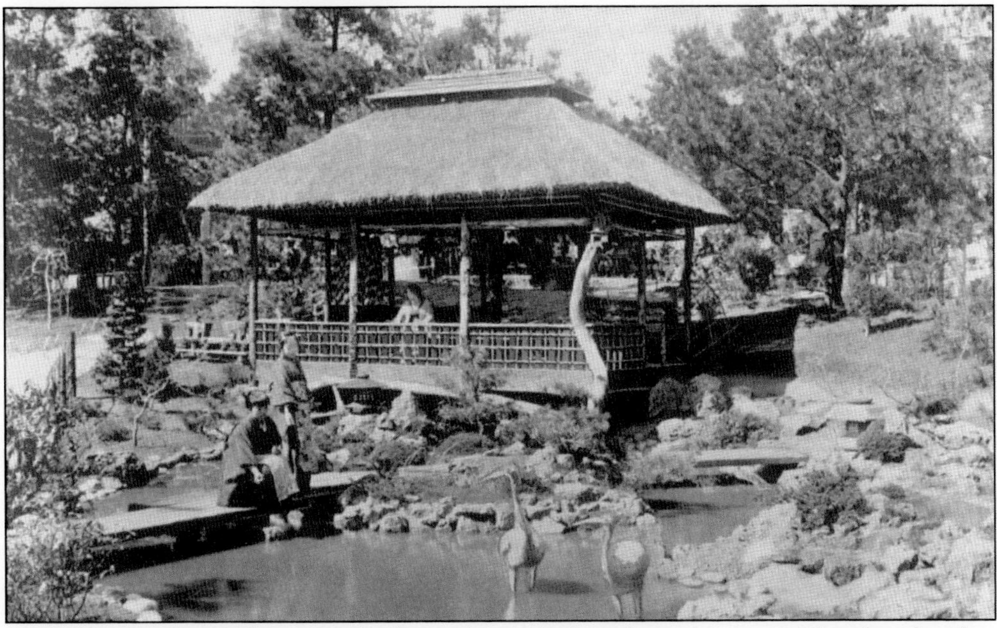

Inside the Japanese Village people shopped, watched juggling acts, or simply relaxed with a plate of cookies and a cup of tea, which cost 10¢. Despite 40 years of trade, most Americans knew little about Japan. The year of the Fair, Lafcadio Hearn published his influential *Glimpses of Japan*, which for the first time accurately presented many aspects of Japanese culture to the rest of the world.

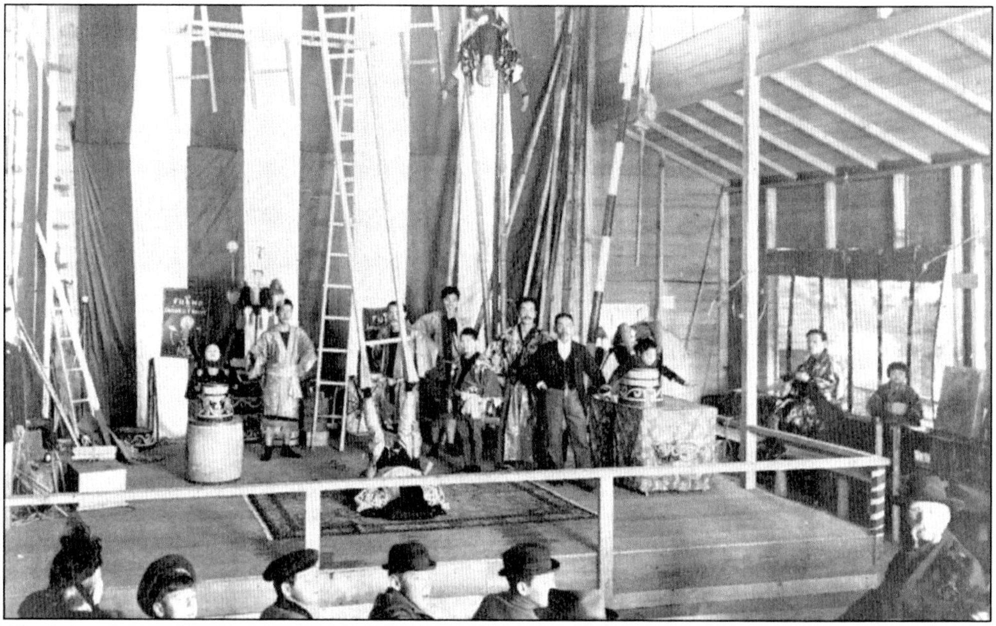

The Theater inside the Japanese Village hosted a troupe of equilibrists who performed balancing acts with consummate skill. Human pyramids were a highlight of the show. Tumblers, wrestlers, and dancing girls added variety. Later in the year, a U.S. Circuit Court ruled that Japanese nationals, being neither white nor black, were ineligible for American citizenship.

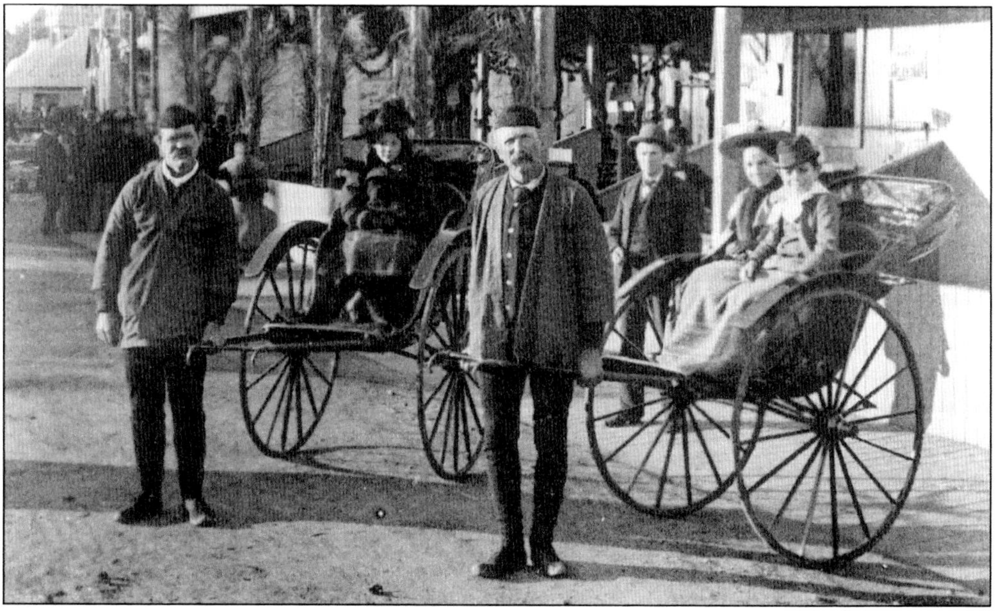

Jinrikishas allowed those with aching feet and ebbing enthusiasm to sit and see the Exposition in relative comfort. The vehicles were authentic, but the attendants were not. Because of the dishonor it brought to their homeland, the Japanese community in San Francisco objected vigorously to local Japanese providing the power; none applied for the jobs. Instead the management hired Germans who painted their faces and donned traditional clothing.

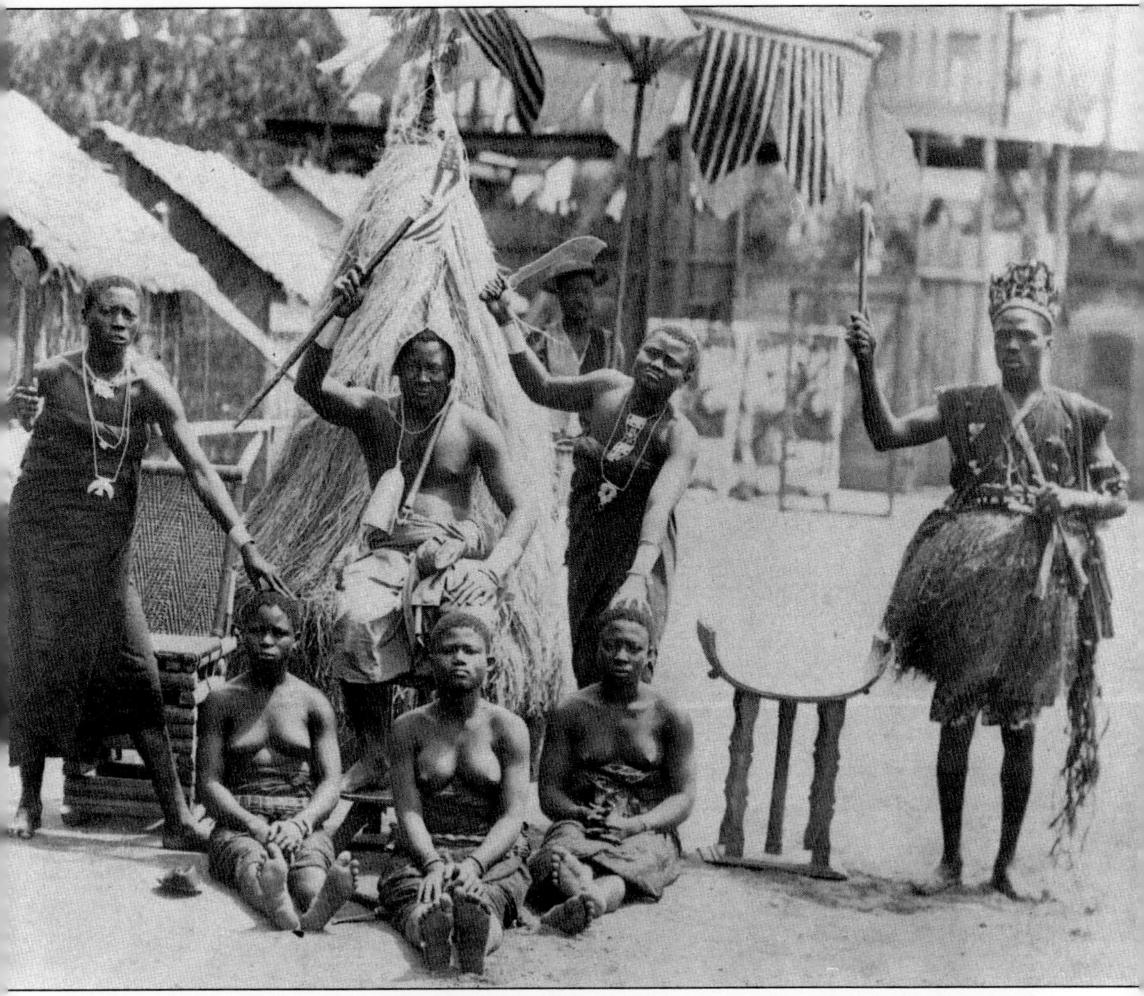

Dahomeyans from West Africa, who enacted traditional war songs and war dances for fascinated fairgoers, also demonstrated their methods of fighting. Some of the women, veterans of the struggle against French colonial rule of their homeland, played leading roles in the mock battles, which surprised many in the crowd. Because the Africans were late in arriving from Chicago, the concessionaire hired San Francisco vaudevillians Bert Williams and Charles Walker to impersonate them by wearing animal skins and stalking through the potted palms. When the group finally appeared, the partners took time to study their culture, which greatly influenced their later work. Two years later, a performance of the cakewalk in *The Gold Bug* made the duo famous. In 1901, with a series of phonograph records for the Victor Company, they became the world's first African-American recording artists. The next year, they used knowledge gained at the Fair to produce and headline *In Dahomey*, perhaps the best known of the turn-of-the-century African-American musicals.

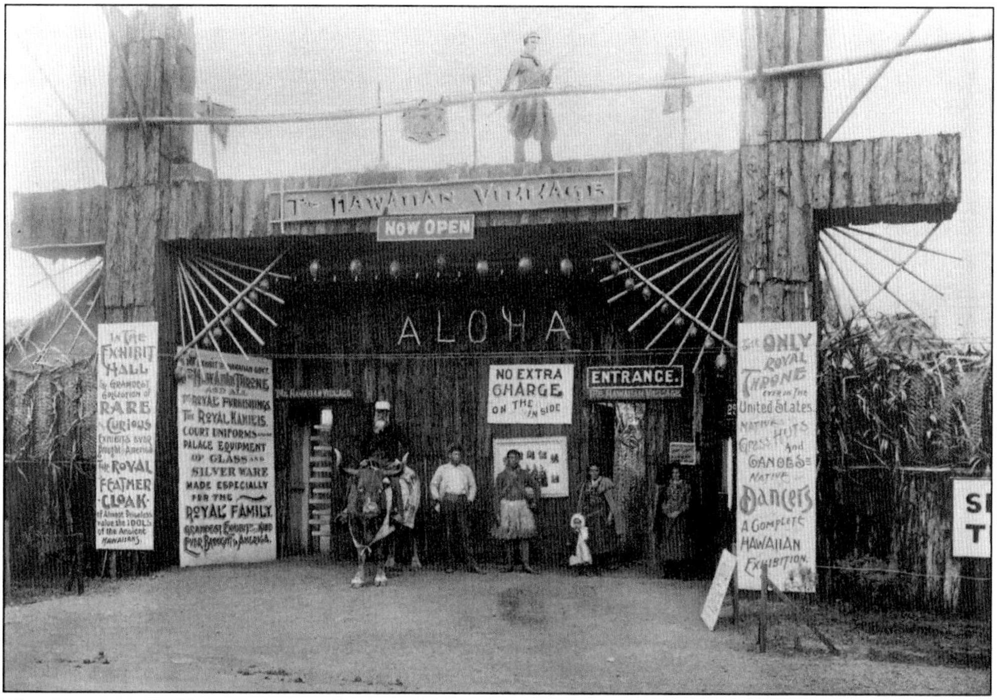

The Hawaiian Village contained numerous attractions, from royal thrones to shark tamers. The facade, an exhibit in itself, was built of native woods and barks, bamboo and palm leaves. The saddle ox added an exotic touch. Admission was 25¢. Regardless of the sign announcing "no additional charges on the inside," hula hula dancers and erupting volcanoes amazed no one for free.

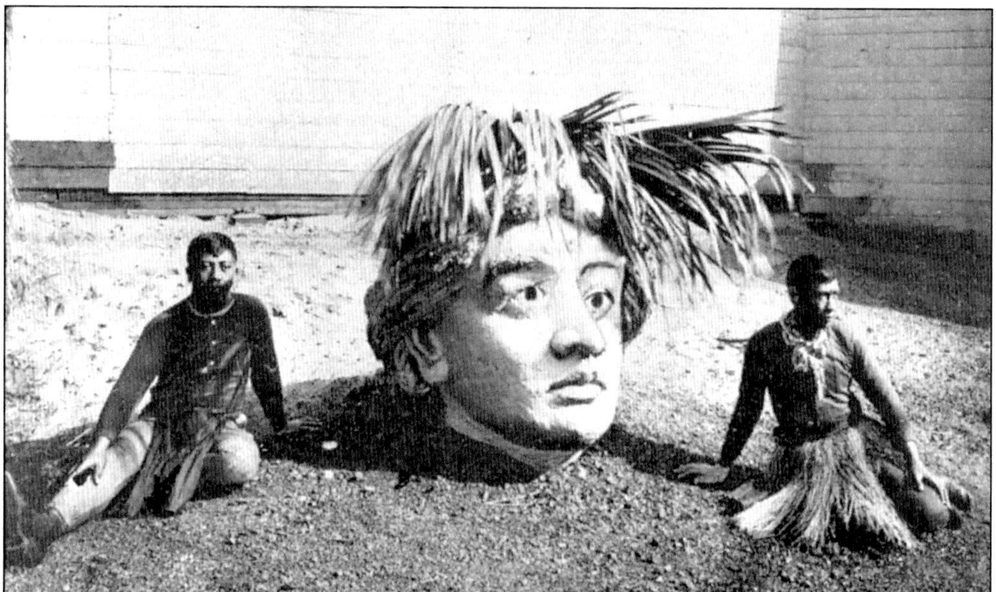

Crowned with palm leafs, the head of Pele, Goddess of Fire, faced the entrance to the Village. Unfortunately, the rest of the statue had been smashed beyond repair on its journey from Chicago to San Francisco. While she was having her bad luck in the U.S., a group of mostly American planters overthrew the rule of Queen Lili`uokalani and established a provisional government.

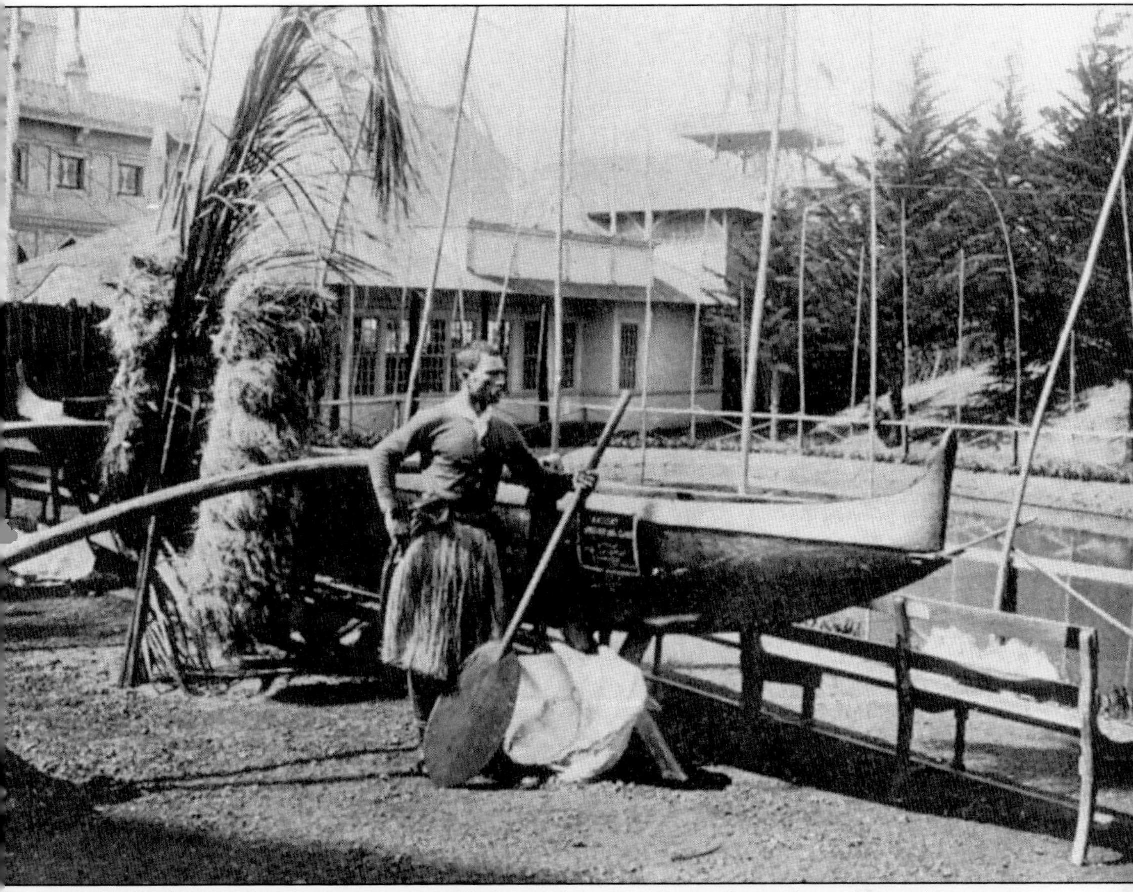

Among the free attractions, visitors watched a shark tamer, shown here, demonstrate how it was done. The Hawaiians were an " entertaining people," according to Martha Greeley, writing in the *Illustrated Pacific States*. She thought they resembled the people of the Chicago exposition's popular Java Village, but added, "These Kanakas, or natives, are larger, more robust, and appear more at home with us than the Javanese." In fact, she found them "so in harmony with Americans I should say it would take very little coercion to cause them to fall into line under the Star-Spangled Banner." Her readers soon found out. On July 4, the day the Fair closed, the provisional government of the Islands declared the Republic of Hawaii, recognized by the U.S. government four weeks later and annexed as an American territory four years after that.

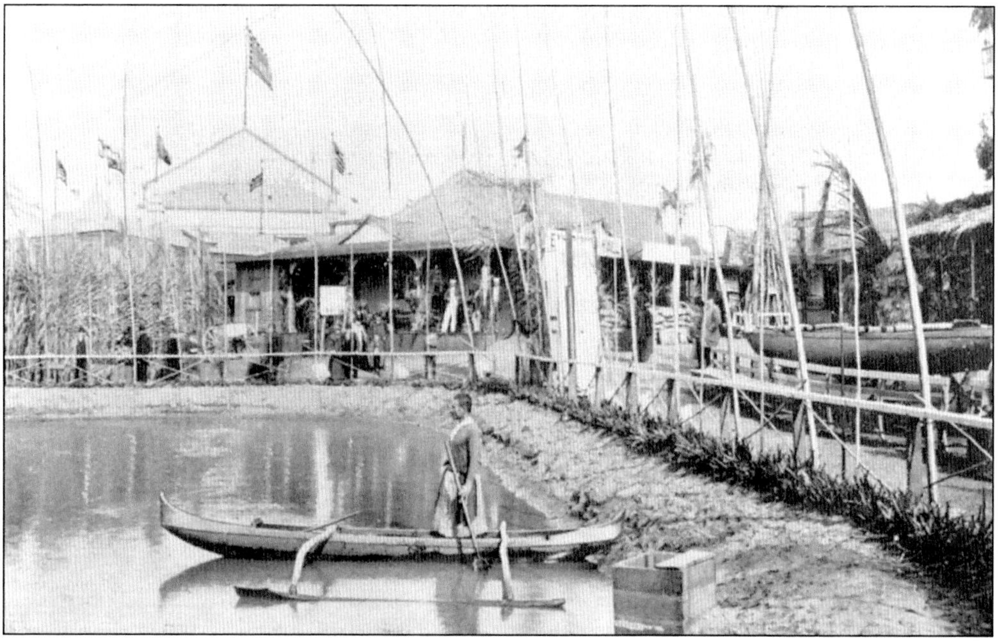

In the Village, patrons saw authentic outrigger canoes, visited a grass house, and browsed a collection of artifacts in the main exhibit hall, all free. A donkey ride cost 10¢ more. To enjoy the hula hula performed by "smiling, dusky young men and maidens fat as seals" required an additional 25¢. Which attraction was most exciting? That depended, at least in part, on one's age and one's wife.

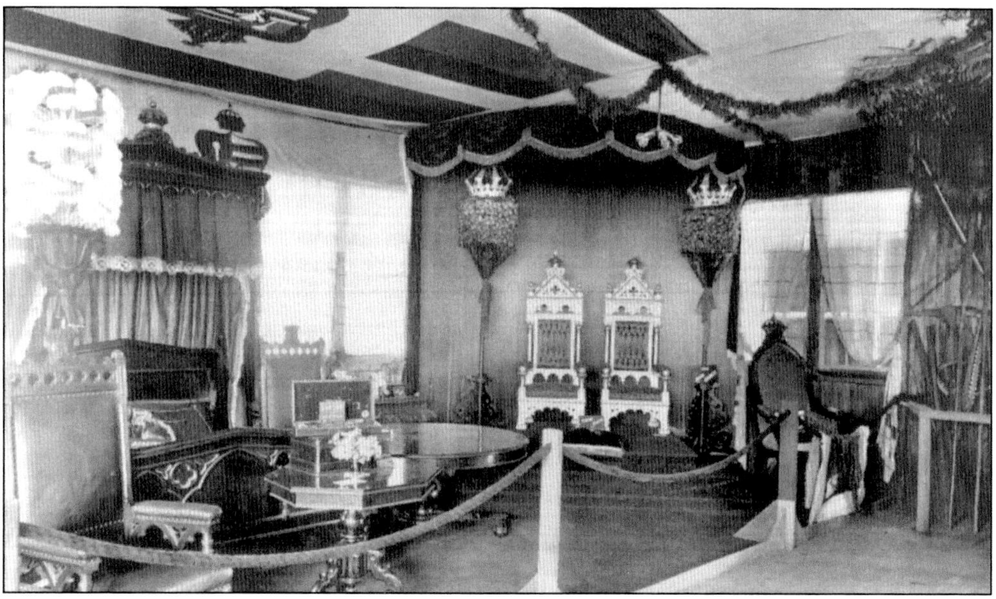

The exhibition building was managed by Mrs. Mary Piilani. Fairgoers saw the thrones of Kings Kamehamehah and Kalakaua, who died in San Francisco in 1891. Lili`uokalani, his sister, succeeded him, but she was deposed in 1893, Hawaii's last monarch. Asked by a visitor, "Why in the world didn't you have the Queen here, too?" Mrs. Cook answered simply, "Because, madam, our Queen is not on exhibition."

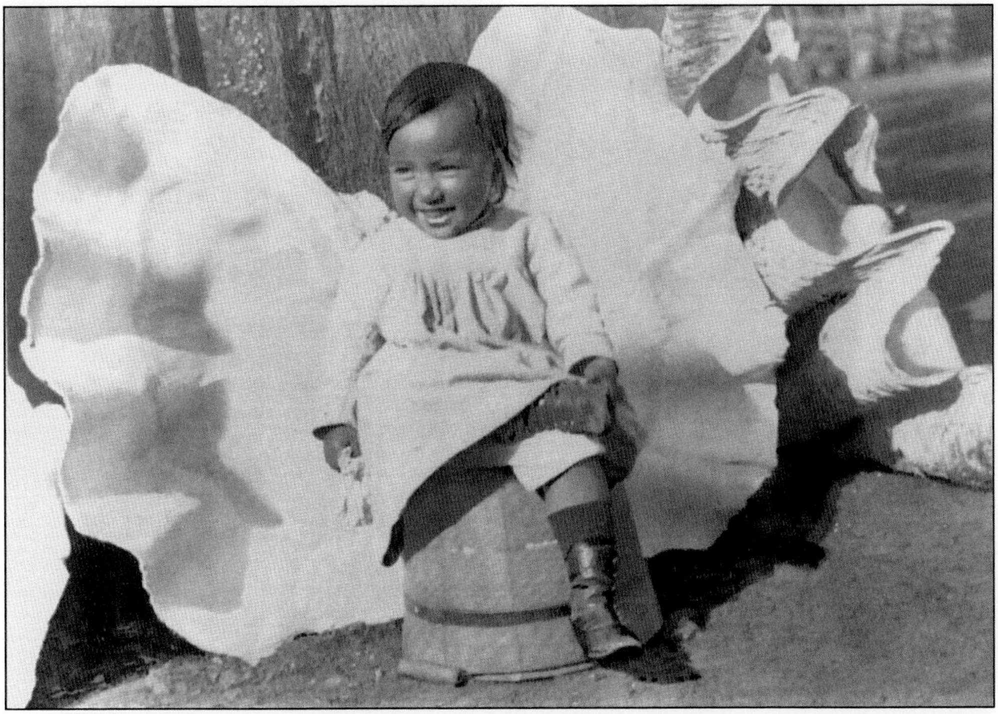

A little girl of the Hawaiian Village poses for a photographer. She sits on an overturned wooden pail among the large abalone shells displayed on the grounds. Her picture won third prize in the amateur photography contest sponsored by the San Francisco *Examiner*.

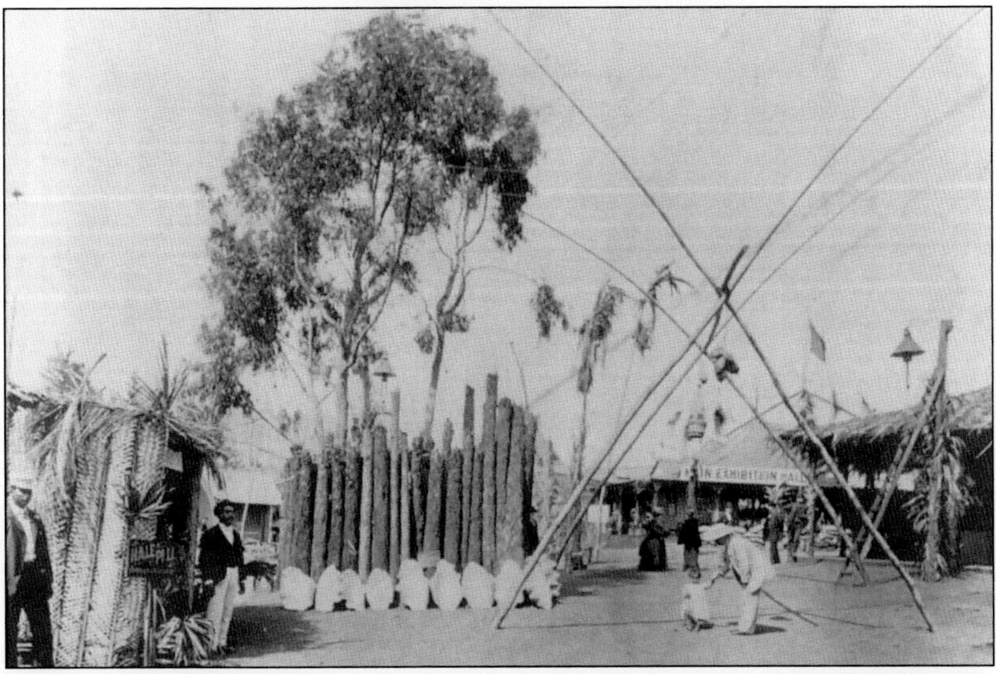

Wandering through the Village, especially with so charming an escort as this little girl, visitors could enjoy all the sights and sounds with clear commentary about each wonder they beheld.

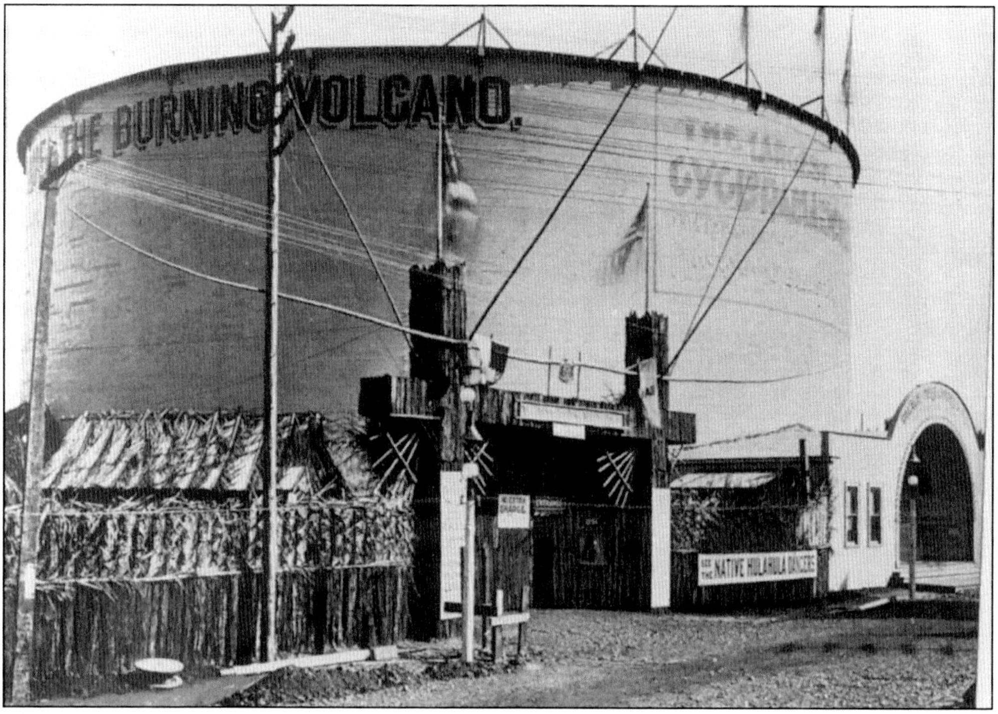

The Cyclorama of Kilauea, 200 feet in diameter, was one of the largest buildings on the Midway. The huge display and its colorful special effects let fairgoers explore the Northern Hemisphere's greatest active volcano in perfect safety and comfort. In an emergency, it could be evacuated in minutes, an advantage not shared with the original.

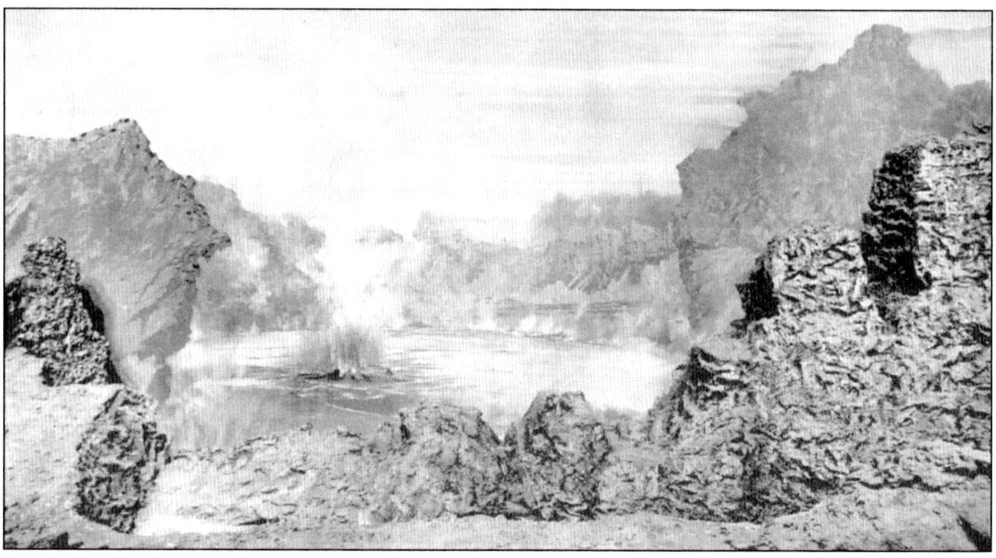

Inside the Cyclorama, fairgoers saw a fierce volcanic eruption, lakes, and streams of burning lava and bottomless chasms. Shifting lights and the shriek of escaping steam added to the realism. The summits of Mauna Kea and Mauna Loa, lower peaks, and a moonlit expanse of the Pacific provided the background. *Overland Monthly* called it "the grandest exhibit at the Midwinter Fair." Such excitement cost only 25¢.

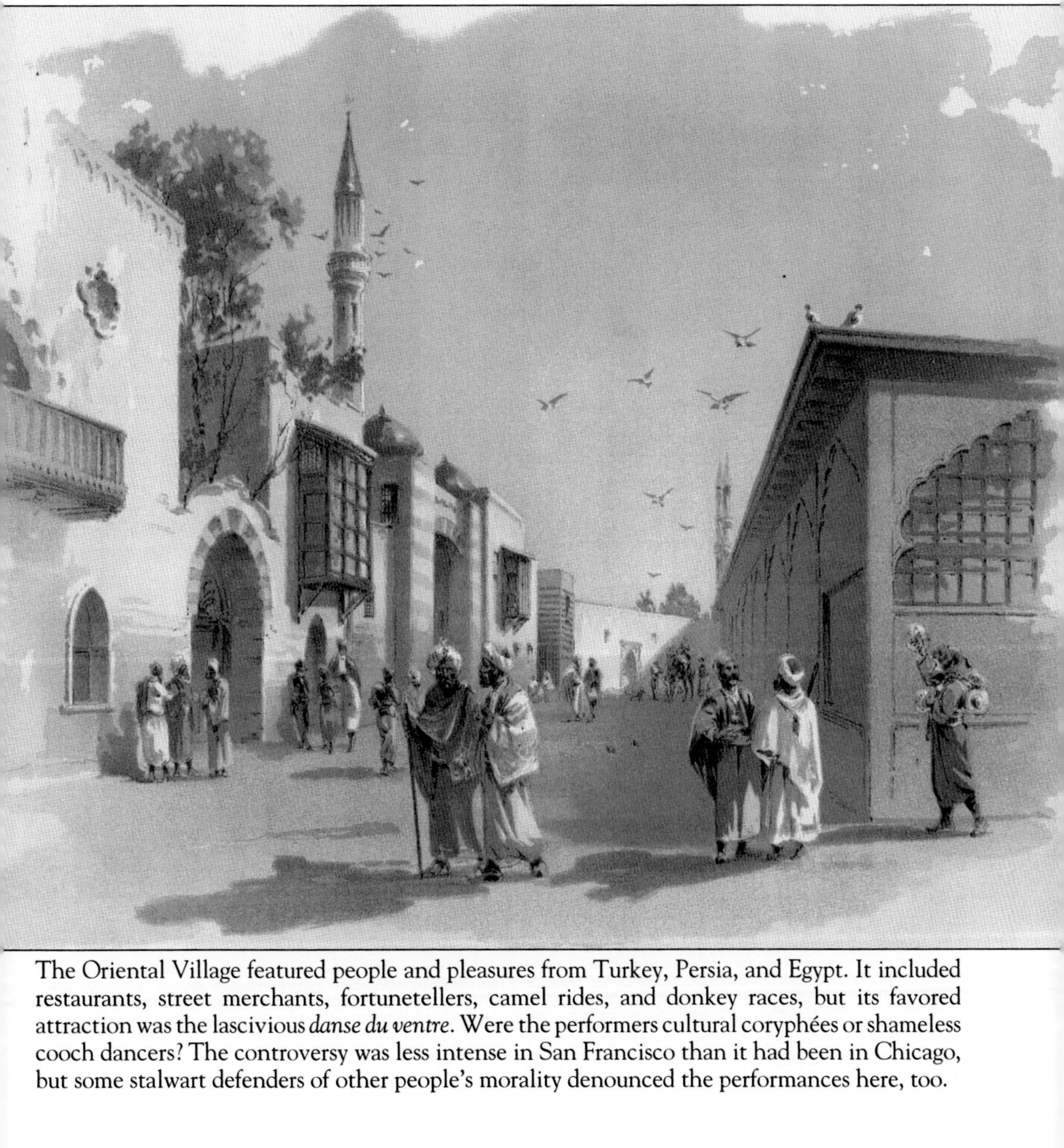

The Oriental Village featured people and pleasures from Turkey, Persia, and Egypt. It included restaurants, street merchants, fortunetellers, camel rides, and donkey races, but its favored attraction was the lascivious *danse du ventre*. Were the performers cultural coryphées or shameless cooch dancers? The controversy was less intense in San Francisco than it had been in Chicago, but some stalwart defenders of other people's morality denounced the performances here, too.

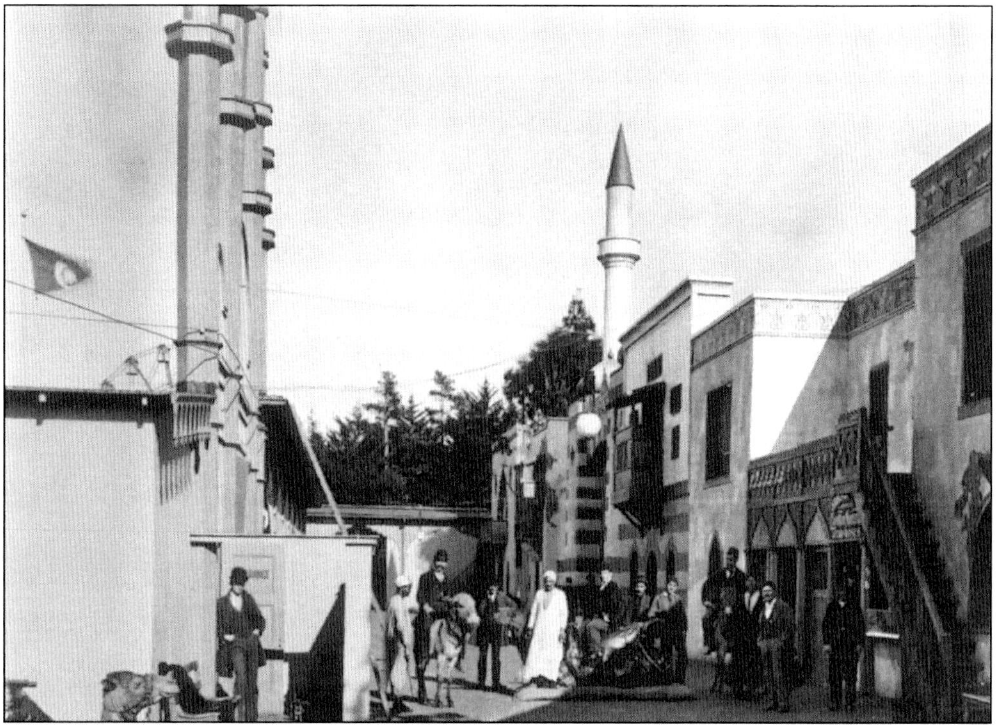

From the booths along Cairo Street, fairgoers purchased beads from Bethlehem, olives from the Mount of Olives, fishnets from Gethsemane, and attar of roses from Samarcand, as well as food and drink. They saw sword fights, candle dances, wedding processions, evening parades, and other celebrations. At all times the street delighted in life: men, women and children, camels, donkeys, dogs, and monkeys—even snakes, when suitably chaperoned.

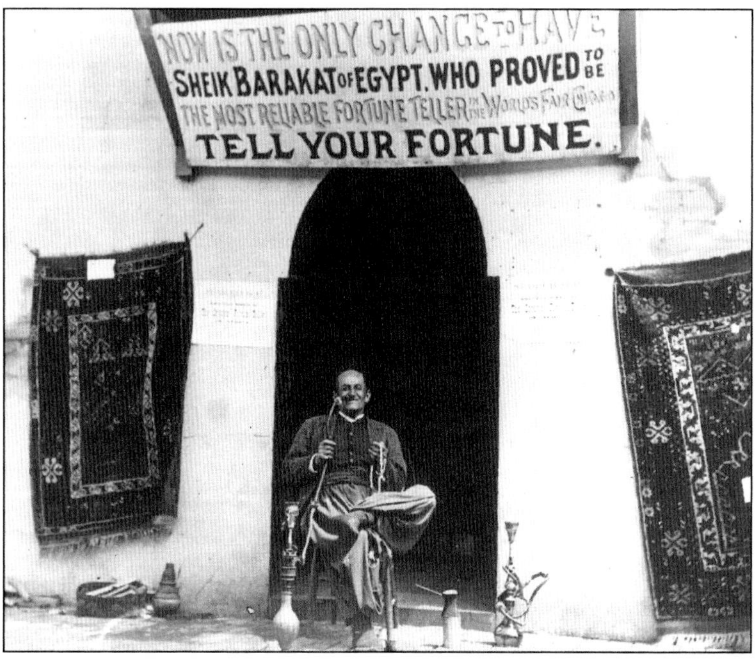

Who wouldn't have wanted a fortune told by Sheik Barakat of Egypt, self-proclaimed as "the most reliable fortune teller of the World's Fair in Chicago?" Certainly he foresaw the photographer's visit to have had himself, his hookahs, and rugs arranged beforehand for the camera.

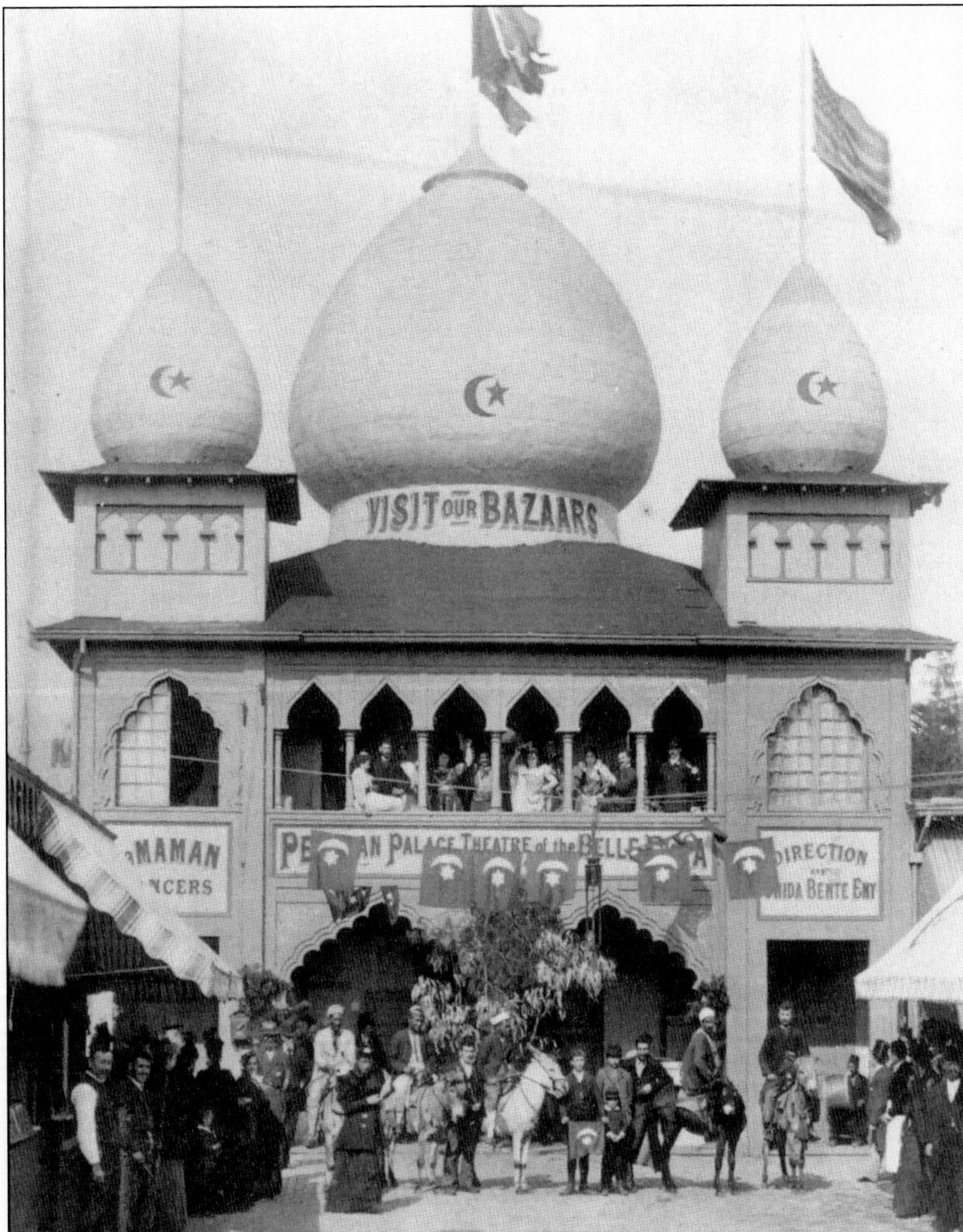

The Oriental Theater was the home of the Belle Baya and other entertainers. Well covered, she poses on the balcony with Ayesha, Soledad, and Fatuna, enchantress of the Nile. Changing mores may make the clucking about their midriffs seem somewhat silly, but at a time when women went swimming fully clothed and some crocheted coverlets for piano legs, even a hint of skin below the neck simply wouldn't do. Inside the Theater, actually called the Peruvian Palace of the Belle Baya, the women of the troupe performed their celebrated belly dance and "fiercely mustached bravos" fought "hot but bloodless combats with scimitars and swords." A similar attraction, featuring Little Egypt, was the scandal of the Chicago World's Fair and appeared at American expositions for years to come.

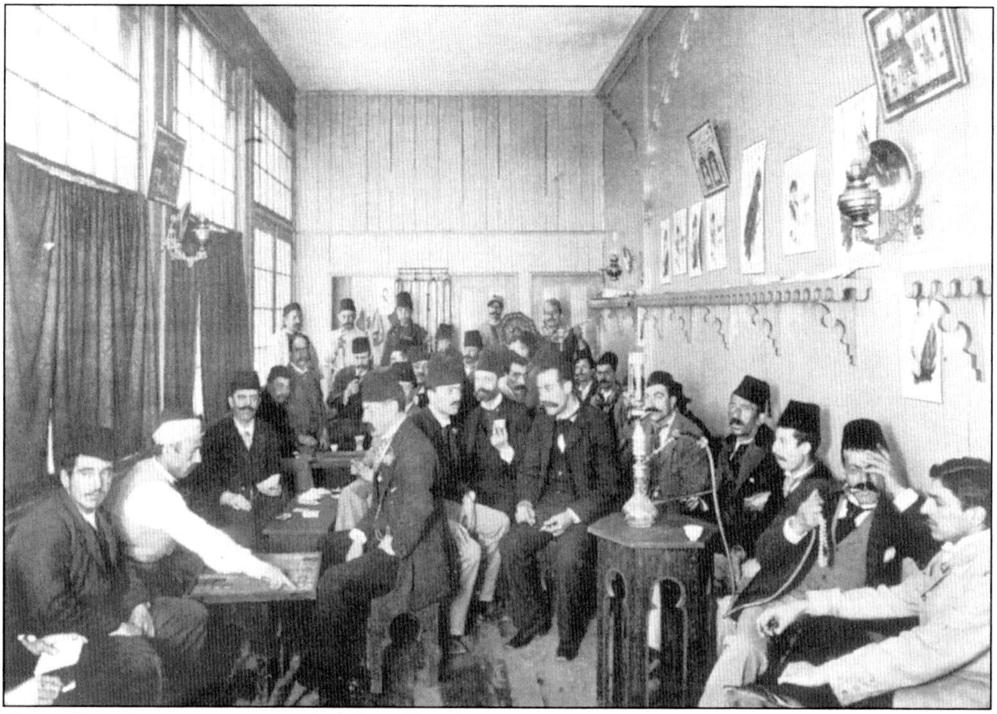

Cairo Street's restaurants and clubs offered the hungry and weary a menu featuring cabbage stuffed with meat and rice, shish kebab, entrees of meat and vegetables baked together, among other dishes. At any time patrons could enjoy a game of backgammon, a water pipe, and a cup of coffee.

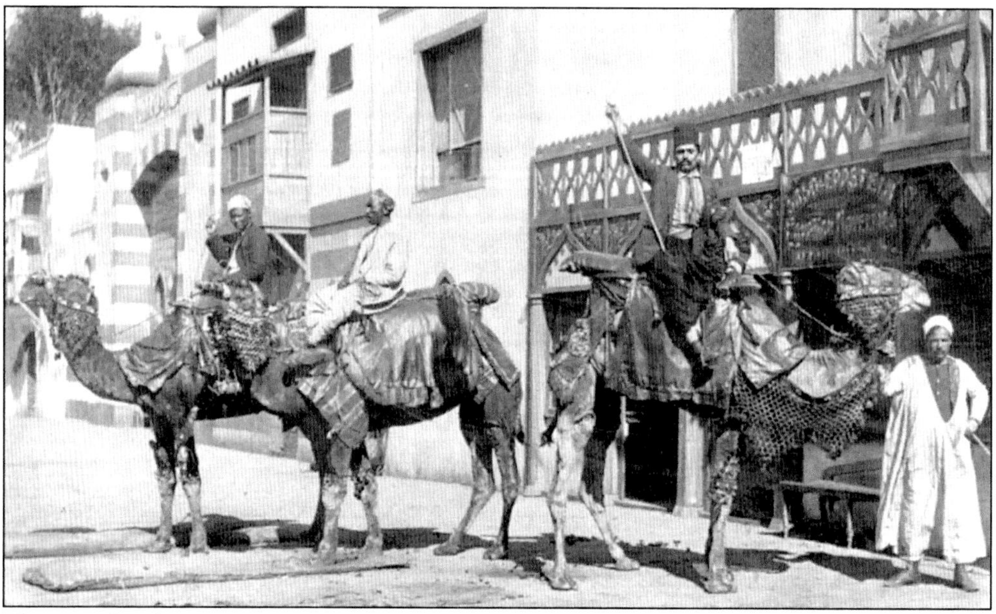

The camels of Cairo Street were named after First Lady Frances Cleveland, actress Mary Anderson, and other prominent women until the management, with a greater sense of propriety than of humor, put a stop to it. A camel ride cost 10¢.

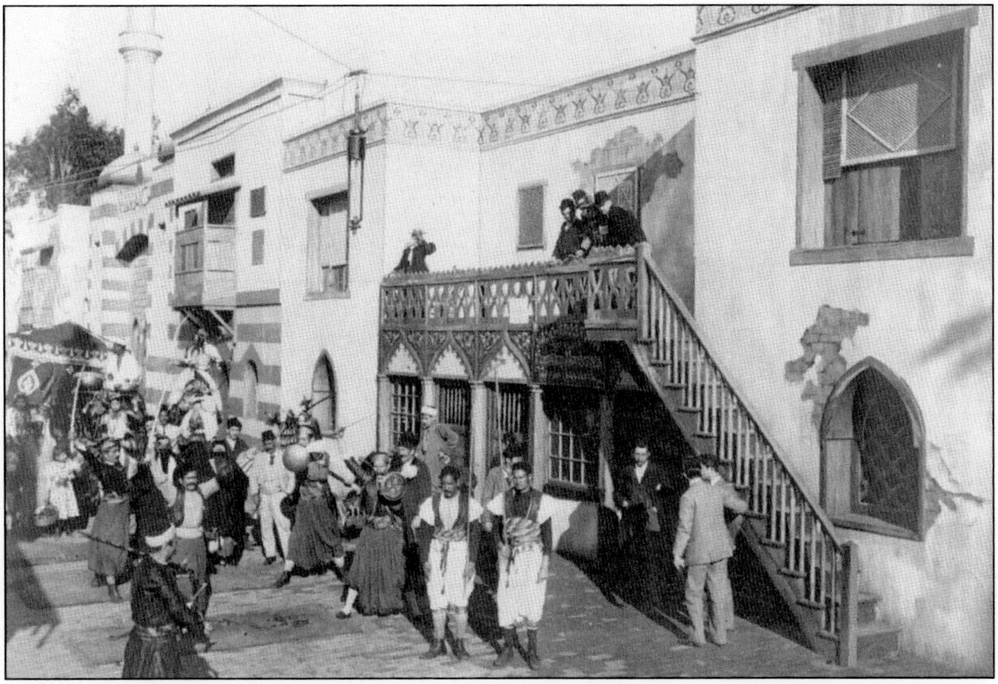

The wedding procession that wended its way down Cairo Street was led by Egyptian swordsmen, followed by musicians, well-wishers, and the newlyweds, who enjoyed their joyful day so much they married almost every afternoon.

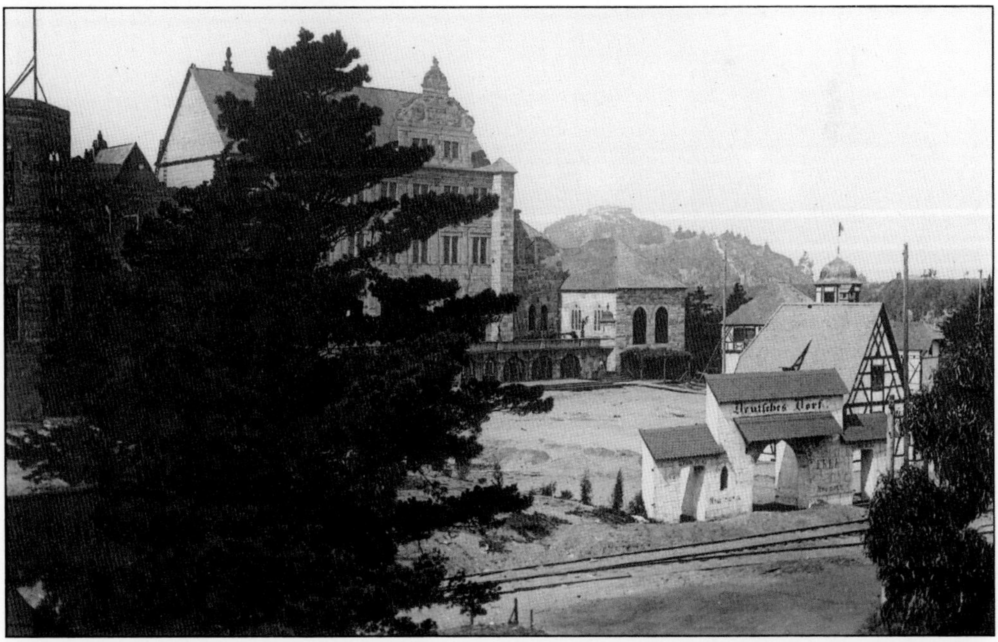

The German Village included a reproduction of Old Heidelberg Castle and the Golden Bear Inn, "where the visitor may take his ease, with his bowl and pipe, and, if his tongue is equal to the emergency, converse with the broad-girthed natives of the Fatherland and forget that he is in California." For some, no doubt, bountiful steins of robust beer produced the same result.

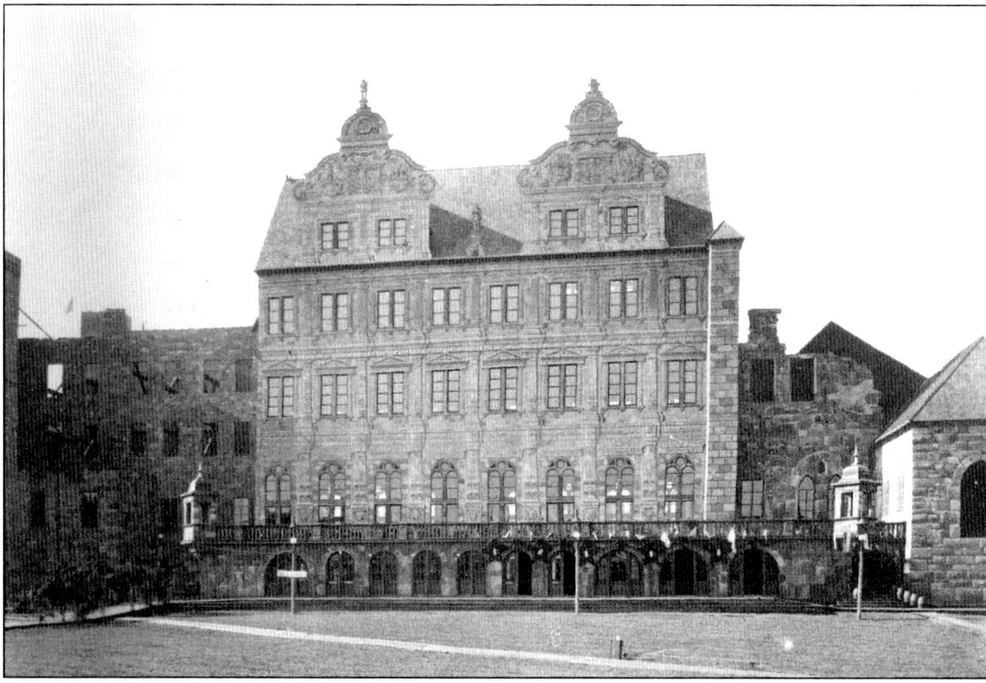

Old Heidelberg Castle, complete with ruins, left, dominated the courtyard, and was the scene of numerous gatherings and band concerts.

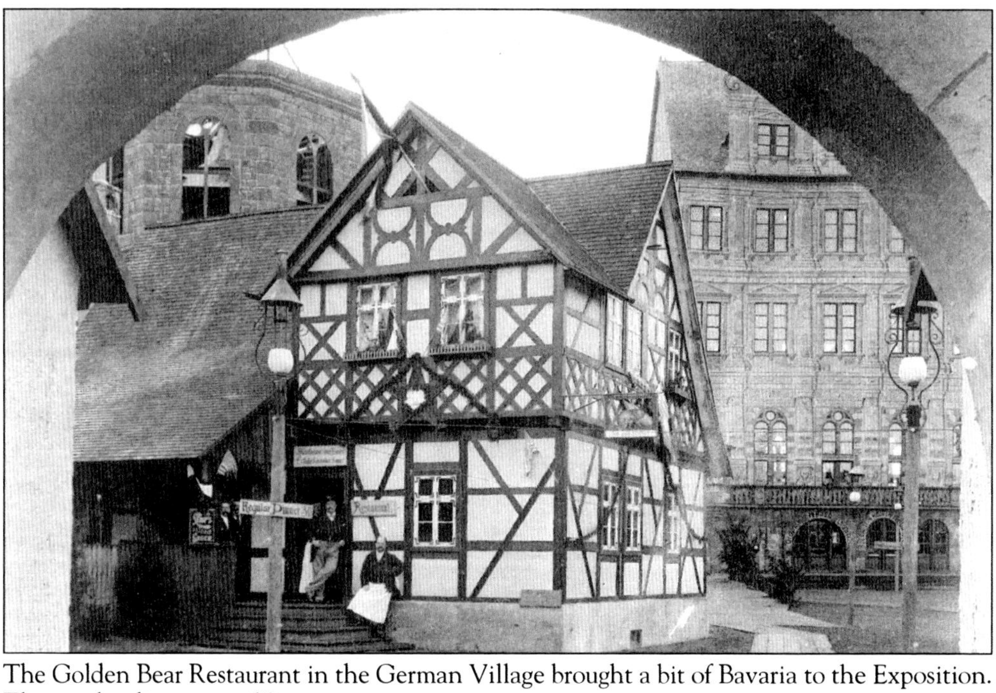

The Golden Bear Restaurant in the German Village brought a bit of Bavaria to the Exposition. The regular dinner cost 30¢.

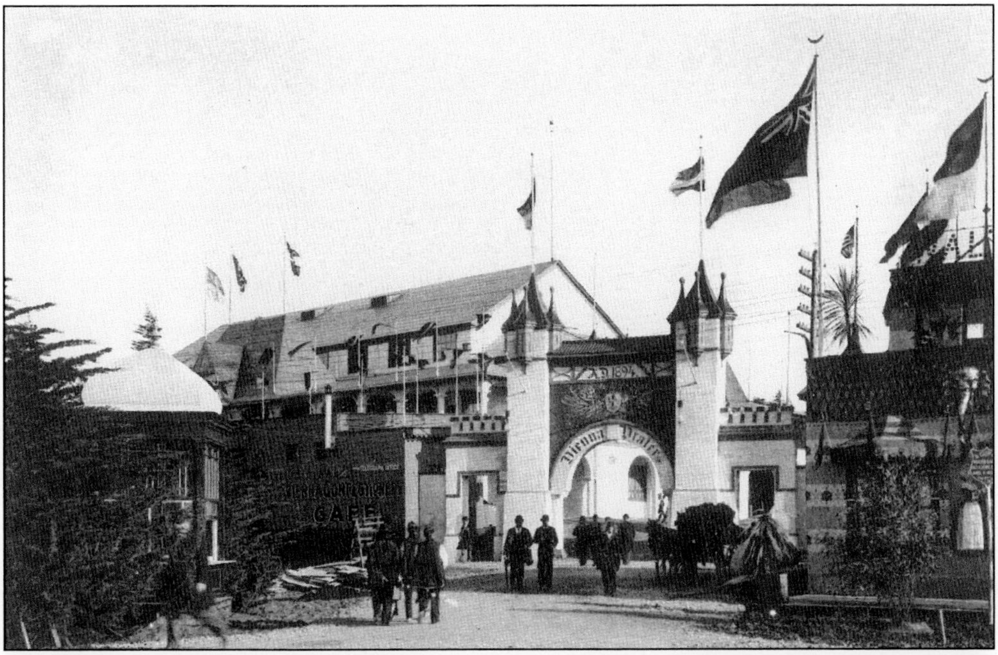

The Vienna Prater contained a restaurant, its own orchestra, and Emperor Franz Joseph Hall, the largest on the fairgrounds, which was used for meetings, banquets, parties, and grand balls.

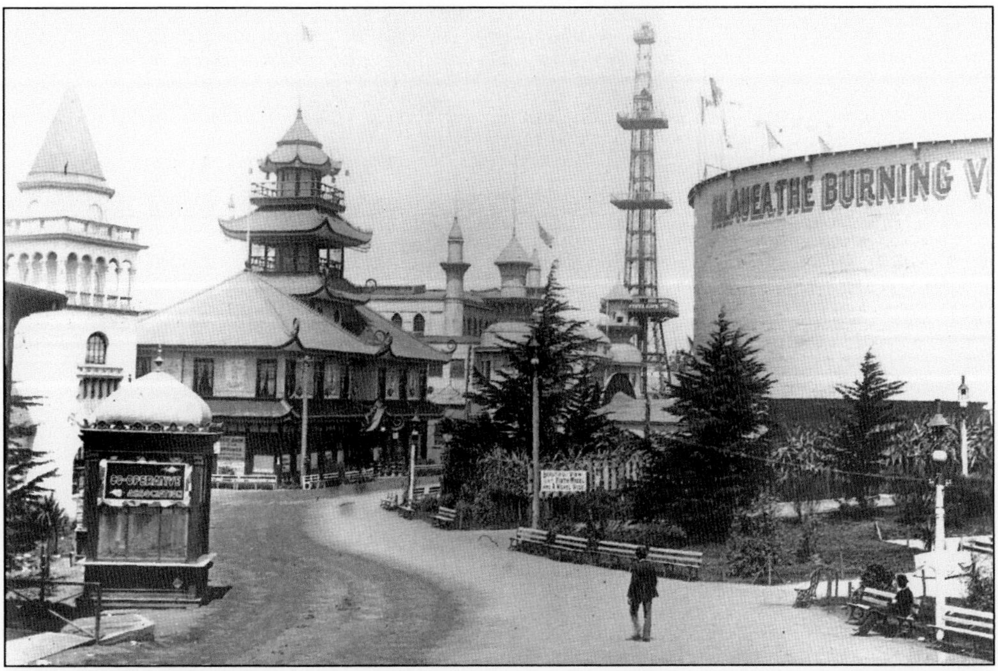

When the Exposition closed, many of the residents of the Villages around the Midway returned to their homes all over the world, but some simply went on to the next exhibition. The Midwinter Fair was the first in a series of great regional expos that took place over the next 20 years in Atlanta, Nashville, Omaha, Buffalo, Charleston, Portland, Hampton Roads, and Seattle. All displayed peoples of the world.

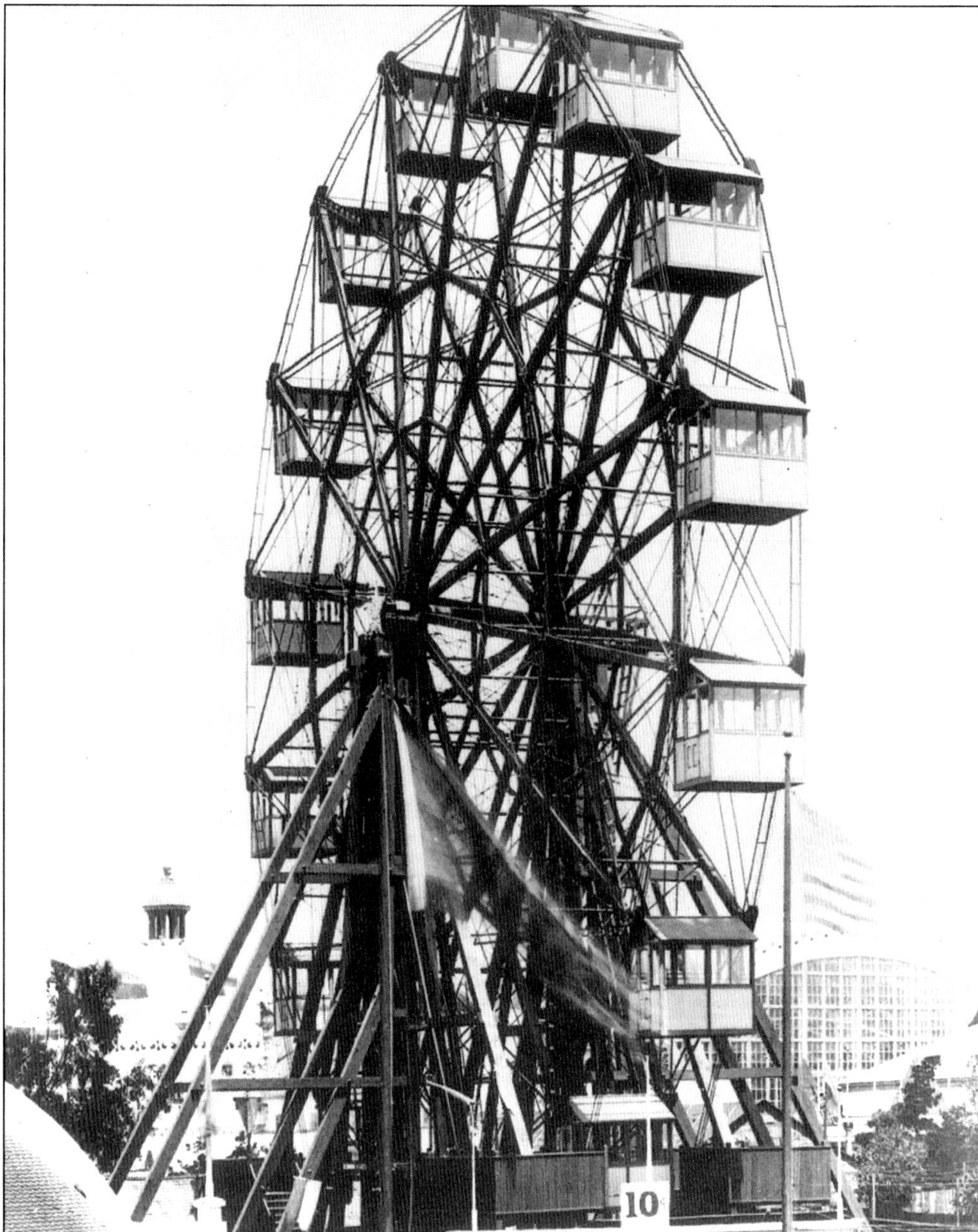

J.K. Firth designed and built this smaller version of the Chicago Exposition's Ferris Wheel which advertised itself as "A Beautiful View and a Novel Ride." Each of its 16 cars carried 10 people to a height of 120 feet above the fairgrounds. That wasn't as high as the Ferris Wheel, but still an honorable altitude for the times, especially in a swaying gondola. Forty horsepower turned the wheel, which, at capacity, weighed 216,000 pounds. A stone and concrete anchorage, weighing more than seven times its super-structure, kept all from toppling over. At the top of the wheel, fairgoers enjoyed an unrestricted panorama of the Golden Gate, the Marin Headlands, and the Pacific Ocean, all for 25¢. The ride, comprising two revolutions of the wheel, lasted about 20 minutes. The attraction hosted the Fair's first marriage on April 1, the bride and groom in one car, their wedding guests in the others.

Four

A Cacophony of Wonders

On the Midway, fairgoers fathomed the depths of Dante's Inferno; walked the streets of the '49 mining camp; saw world of the Camera Obscura; outwitted the apparitions of the Haunted Swing; raced thoroughbreds at the automatic race course; and visited an ostrich farm, a captive balloon, a ferocious lion, and much more.

Tickets to exotic places or heroic adventures typically cost 10¢, occasionally 25¢. "Shouters" or "spielers" described the wonders within their singular concessions with a facility that could persuade pinchpennys to pry open their pocketbooks. Fortunate fairgoers who required only a glimpse of the extraordinary found free previews of the goings-on in front of some attractions, but that was hardly the same as experiencing it all. Wanting to welcome patrons often, some enterprises offered many amazements, each at additional expense, of course.

The Fair's biggest controversy occurred on the Midway when Colonel E. Daniel Boone announced a one-time-only wrestling match at his Animal Arena between Parnell the Lion and Siskiyou the Bear. All 1,500 tickets sold at the considerable sums of $10, $15, and $20 each. The park commissioners then stated they would close the Exposition if the encounter took place, compelling the Fair's directors to urge its cancellation. By fight night Boone's entire establishment was in the hands of the San Francisco SPCA. With the help of 20 Midwinter Fair guards and 40 San Francisco police, there was no fighting inside or outside the arena. Boone got his animals back, but the contest stayed cancelled.

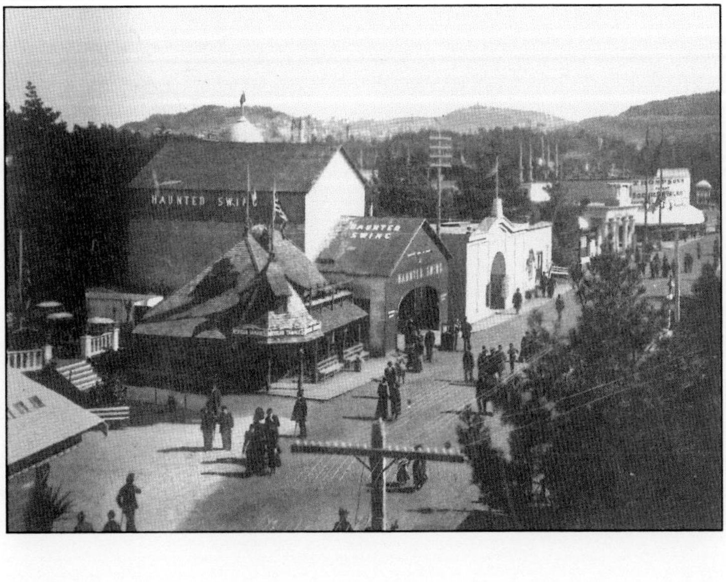

The Midway along South Drive included the Tomales Cottage, the Haunted Swing, the entrance to the Oriental Village, and the boarding station for the Scenic Railway. Fairgoers liked them all. Across the street they found the Arizona Museum, the Alameda County Building, and Boone's Animal Arena. Additional adventures awaited them on North Drive.

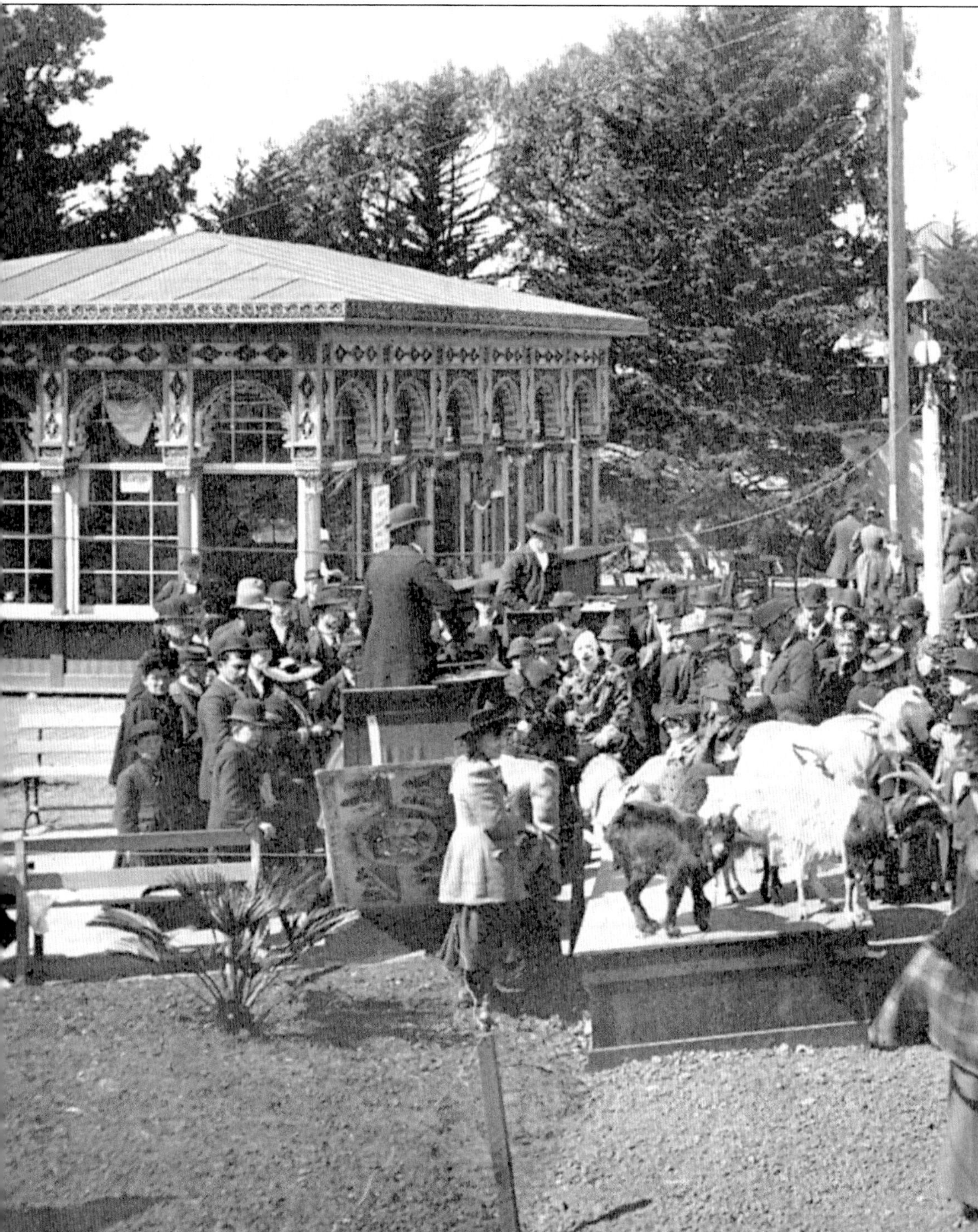

After the burden of browsing all morning amid the main exhibits halls' rocks and rhododendrons, girls and boys of every age willingly went off to explore the Midway. Could this youngster have been more enchanted by the charming miss behind him than the tantalizing attractions before

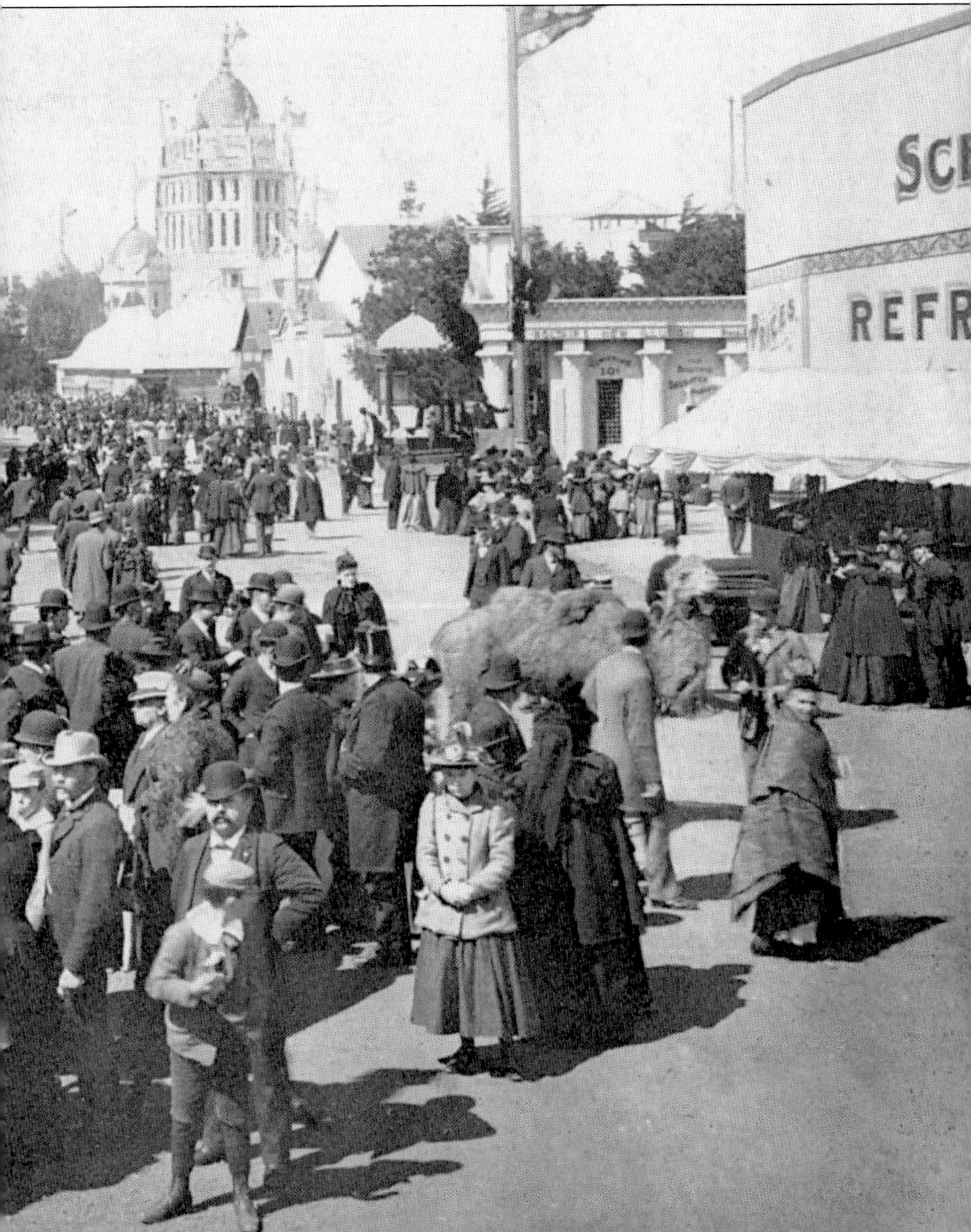

him? Perhaps the ambulant camel behind her has captured his attention from the kaleidoscopic clown and wondrous woolies of Boone's Animal Arena to his right.

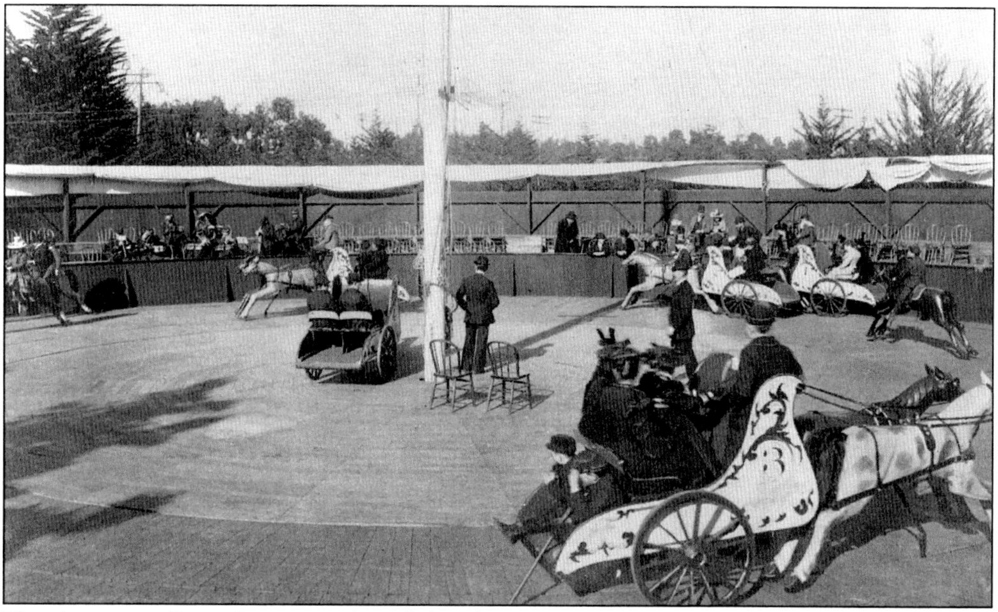

The Automatic Race Course combined the pleasures of a merry-go-round with the excitement of a sports competition. Horses and chariots connected to ropes and pulleys under the platform that moved them around the track. Everyone began noses even at the starting post, but the fix was in. The attendant could slow down or speed up whomever he chose. The ride, a children's favorite, apparently was an adult's favorite, too.

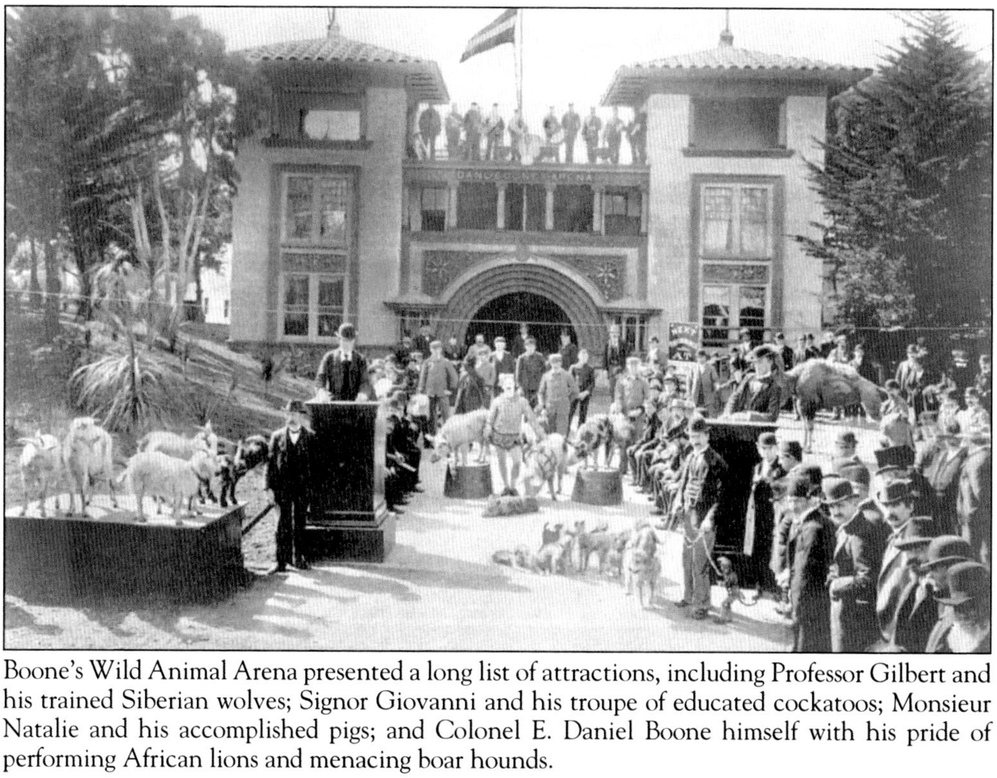

Boone's Wild Animal Arena presented a long list of attractions, including Professor Gilbert and his trained Siberian wolves; Signor Giovanni and his troupe of educated cockatoos; Monsieur Natalie and his accomplished pigs; and Colonel E. Daniel Boone himself with his pride of performing African lions and menacing boar hounds.

Dancing bears performed outside Boone's Wild Animal Arena as a preview of the show inside. From time to time, passersby could see monkeys, goats, dogs, and birds doing bits of their acts in the courtyard.

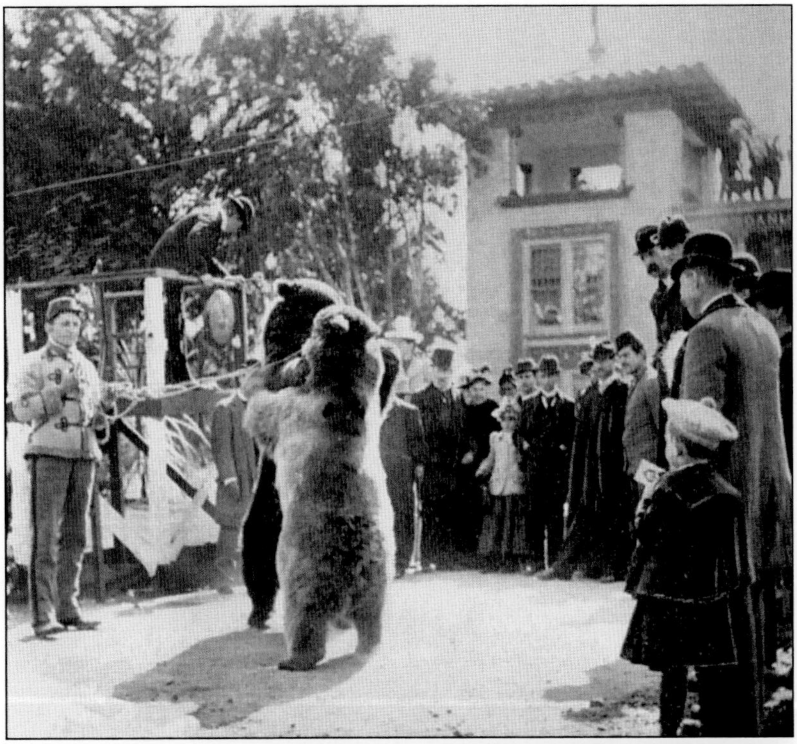

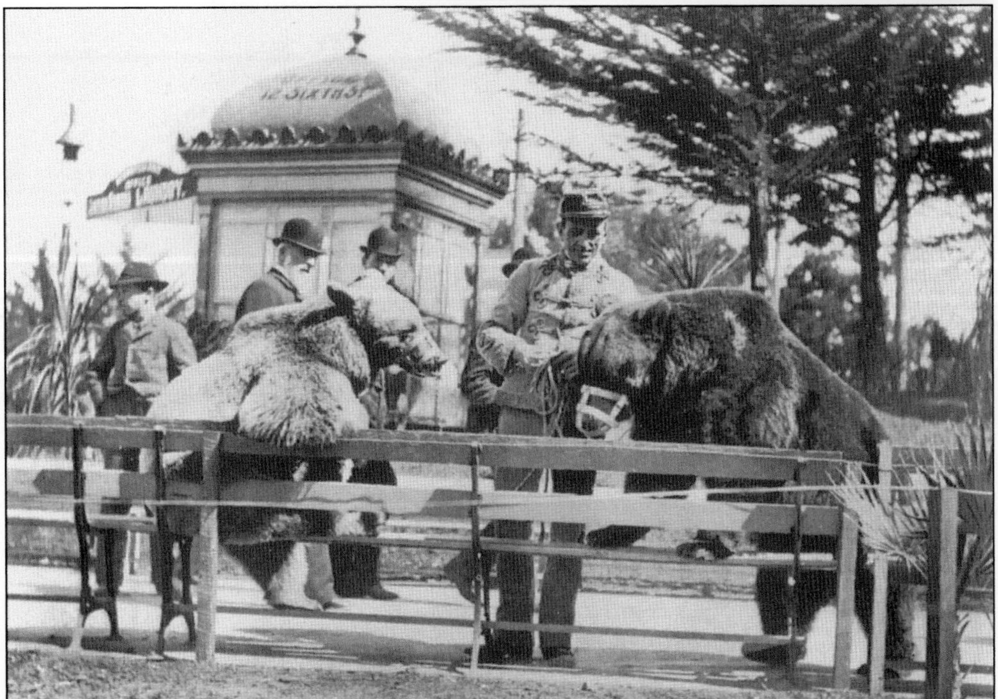

Between performances, some of the bears enjoyed a little sunshine and conversation on an inviting park bench, just like any grizzled visitor after a busy morning at the Fair. This snapshot won second prize in the amateur photography contest sponsored by the San Francisco *Examiner*.

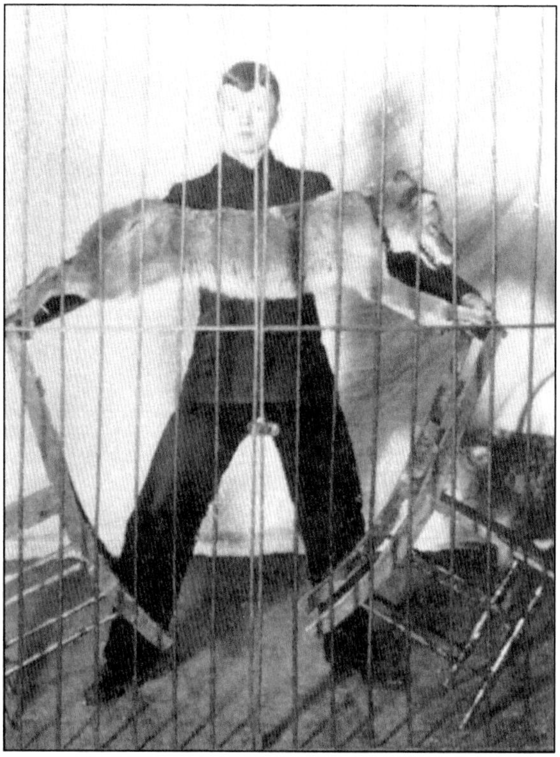

Fairgoers delighted in the tricks the animals performed. At 50¢, the show was the most expensive attraction on the Midway, but it lasted 90 minutes.

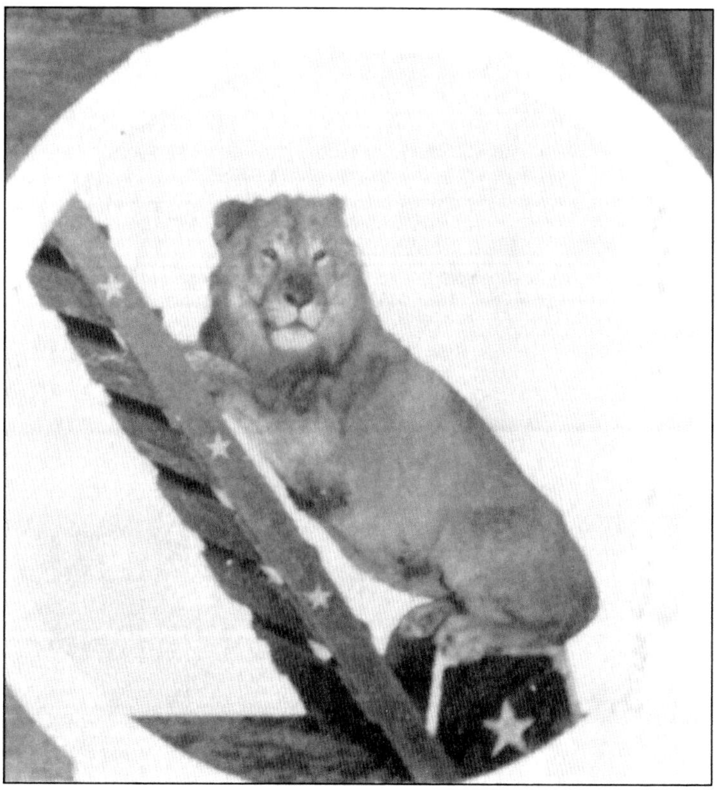

Parnell the Lion was the undisputed star of the Wild Animal Arena, partly because of his declared ferociousness—we shudder to call it his animal magnetism—but especially because he killed a lion tamer during the first month of the Fair. Too valuable to be destroyed, he was back at work the next day, "his popularity among sight-seers," according to the San Francisco *Examiner*, "increased a hundred fold."

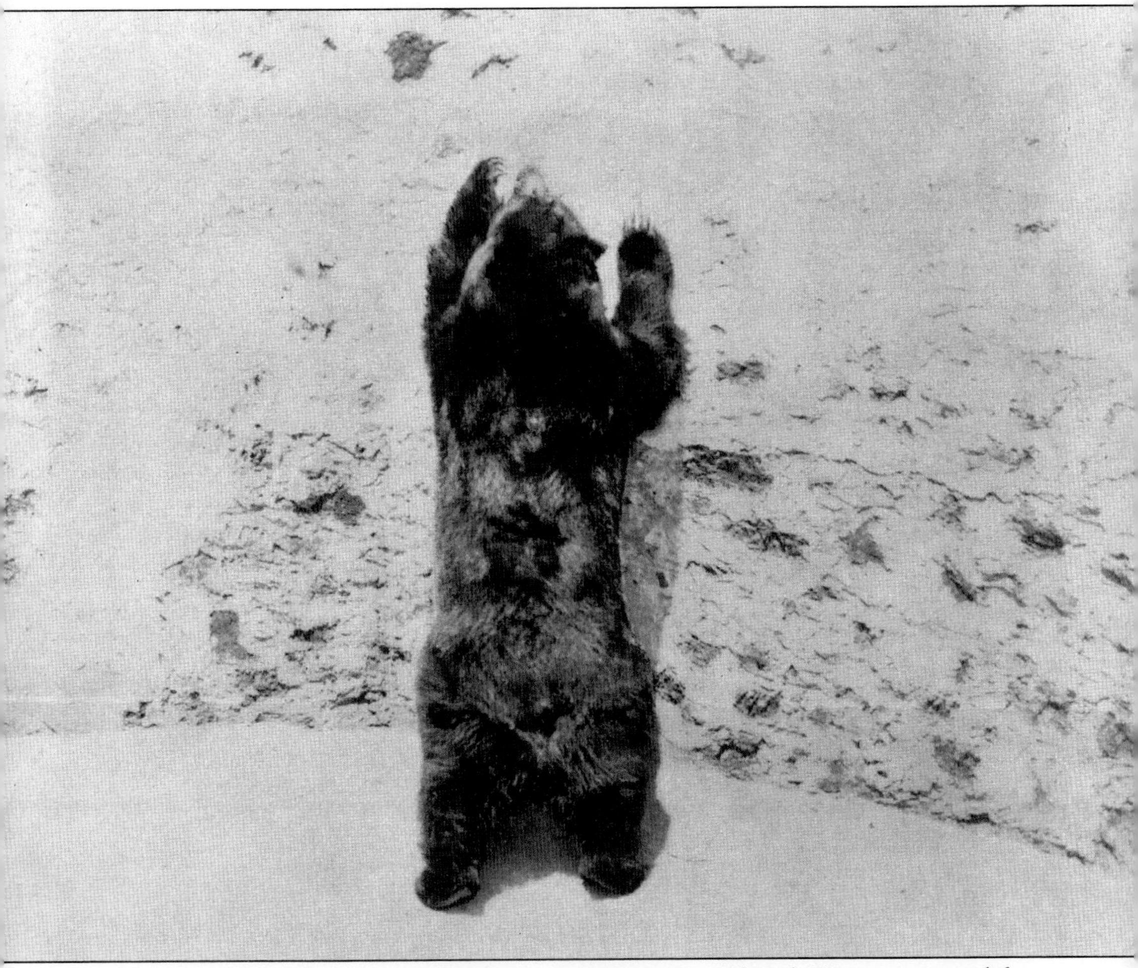

Monarch the Bear fared better, if long life in captivity were preferred. He was captured for William Randolph Hearst as a gift to the San Francisco zoo, which declined, although his descendants may be found there today. He went instead to live at a local amusement park, and then moved to Golden Gate Park for the Fair; his home, shown here, was a huge concrete pit. When the Exposition closed, he relocated to an iron cage in a different part of the park, where he remained until his death in 1911. Twenty five years later he appeared on coins issued by the U.S. Mint to commemorate the opening of the Bay Bridge, the first bear so honored. Monarch himself, carefully conserved, may be seen in the California Academy of Science, not far from his final home. He was the last known California Grizzly.

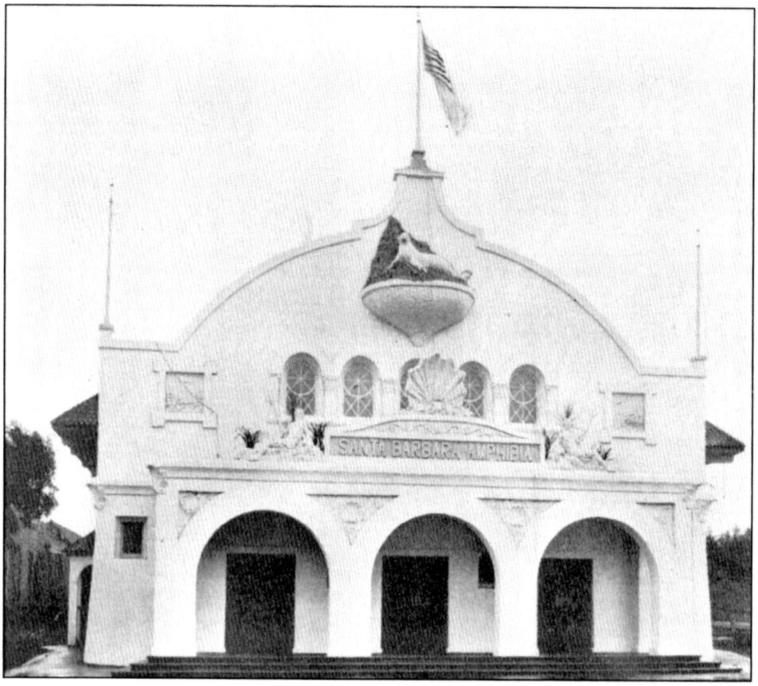

The Santa Barbara Amphibia, situated next to the Firth Wheel, was the first Exposition building to be completed, at which time it opened for business. It contained a colony of 15 sea lions and sea tigers. On sunny days, many early arriving fairgoers were startled and amused to find the beasts, dripping and tottering, out for a morning's amble around the Midway.

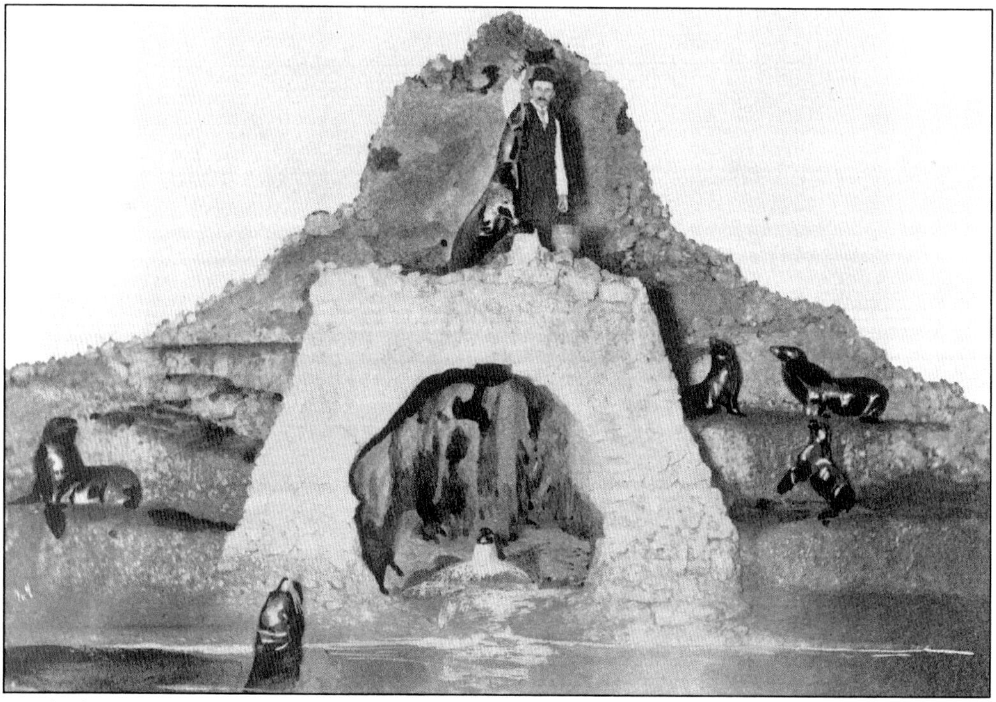

Inside the Amphibia the audience watched the animals' attendant "demonstrate the power of man over the monsters of the sea," mostly by tempting them with herring. For these tantalizing tidbits the seals climbed steps, pushed open a wicket gate with their snouts, then jumped on a chair and bounded up its back. One learned to fire a pistol by pulling a string attached to the trigger; accuracy was not paramount.

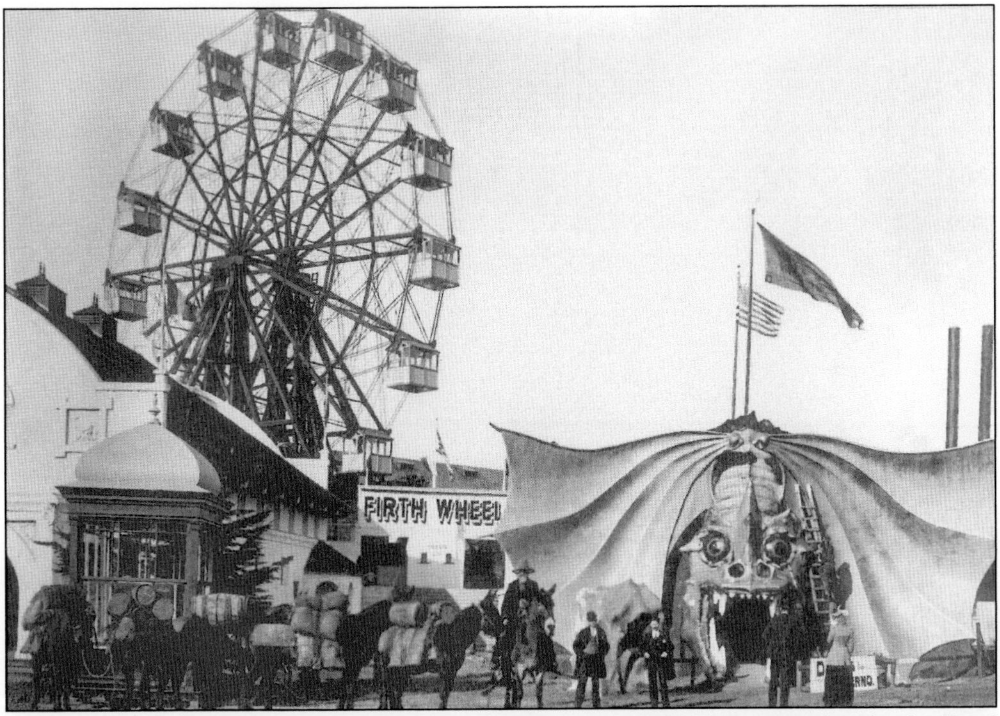

Dante's Inferno, on South Drive near the Santa Barbara Amphibia and the Firth Wheel, offered a look at the afterlife of sinful mortals for 25¢. Patrons entered through the demon's head. Inside they saw burning lakes, dancing skeletons, and bottomless pits, the presumed fate of all fairgoers of incautious biography.

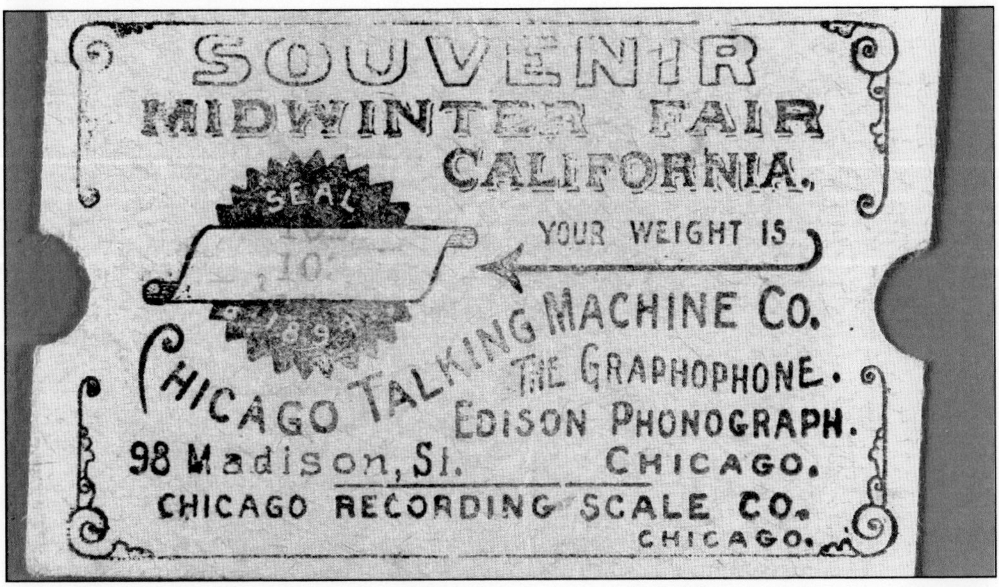

Depending upon a fairgoer's tastes and disposition, one of the Midway's scarier attractions was the recording scale. Was a weight of 103 pounds welcome news or bad? "You were born under a lucky star," says the fortune on the back, so likely the enlightenment was well received. This tiny messenger measures 2 by 1.25 inches.

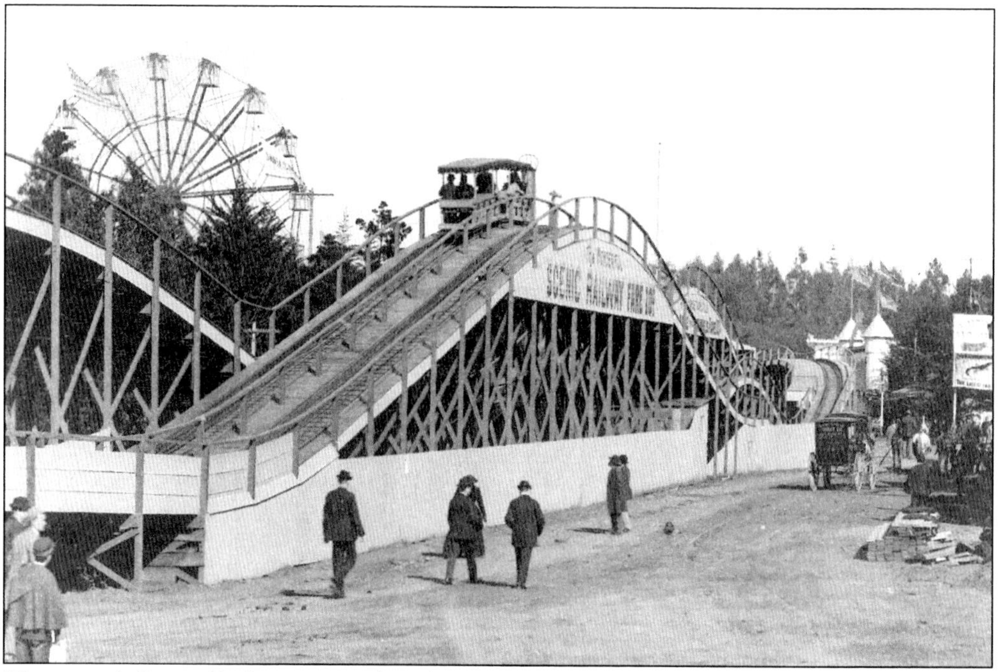

Thompson's Scenic Railway gave riders the exhilaration of a roller coaster and the pleasure of a sightseeing excursion. They ascended by cable car and descended by gravity, looking out at the fairgrounds before entering a forbidding cavern. Some 500,000 passengers took the trip at 10¢ each. No one paid with an 1894-S dime, however. The San Francisco Mint at 5th and Mission Streets produced only 24 that year, none for circulation.

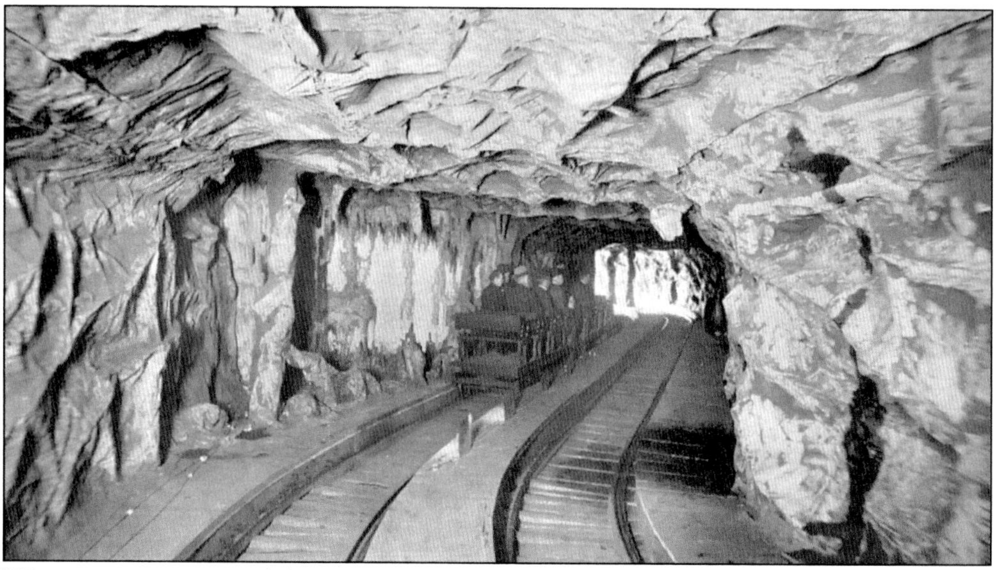

When passengers entered the cavern on the Scenic Railway, they immediately plunged into total darkness. Then, suddenly, bursts of light brought their surroundings to life. Designed to invoke the tunnels built for the railroads through mountains and under rivers, many fairgoers instead thought they were travelling into deep mines or caves. According to the Chicago *Daily News*, "The journey, instead of being *frightful*, is *simply delightful*."

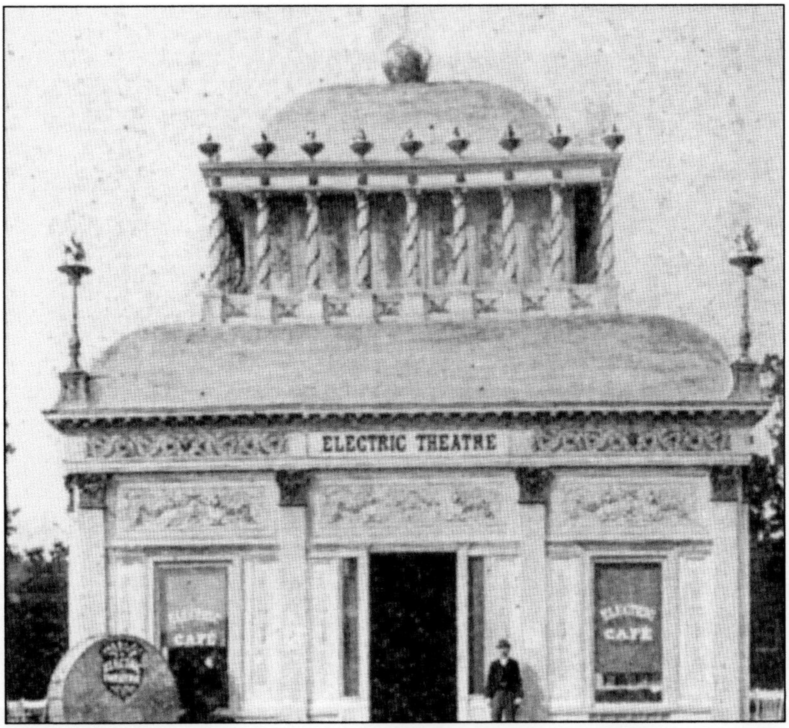

The Electric Scenic Theater allowed spectators to experience a day in the Alps, complete with an obligatory violent afternoon thunderstorm. Live instrumental music accompanied the changing conditions. Traveling in comfort to Switzerland from San Francisco would have cost audience members many hundreds of dollars and taken almost two weeks. Here they spent 25¢ and 30 minutes. Before or after their arduous adventure, sightseers could refresh themselves at the Electric Cafe.

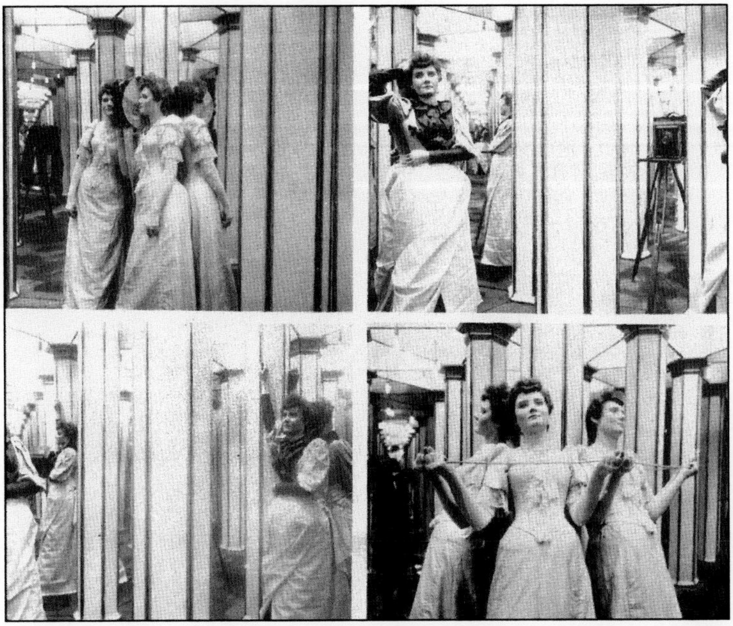

The Mystic Mirror Maze so baffled visitors that one of the fads of the Exposition became taking uninitiated friends through it to watch their confusion and wonder. Only one woman posed for the photographer, visible in the mirrors, but the optical illusion made it seem that there were more. The attraction went next to Sutro Heights, then to Playland at the Beach, where it amused patrons until the 1960s.

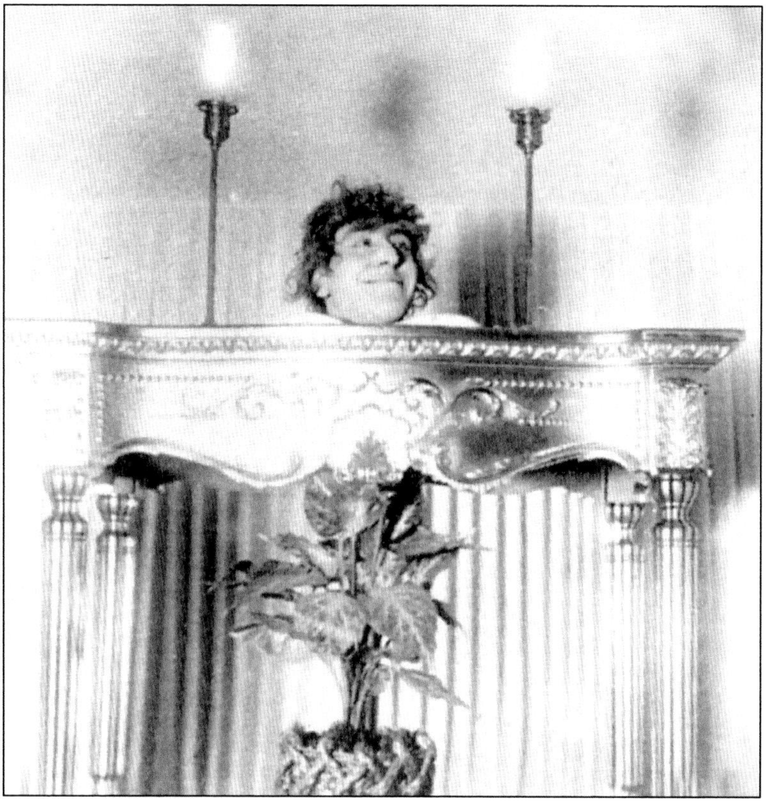

Other illusions of the Midway included Rolla, whose body from the waist up rested on a fixed stand; Amphrodite, who apparently floated in mid-air; the Maid of the Mist, who emerged from a very small fountain, then disappeared back into it; and the Sphinx's Table, shown here, upon which perched a man's head, talking and telling jokes so hoary that they barely got a wheeze even in grandfather's day.

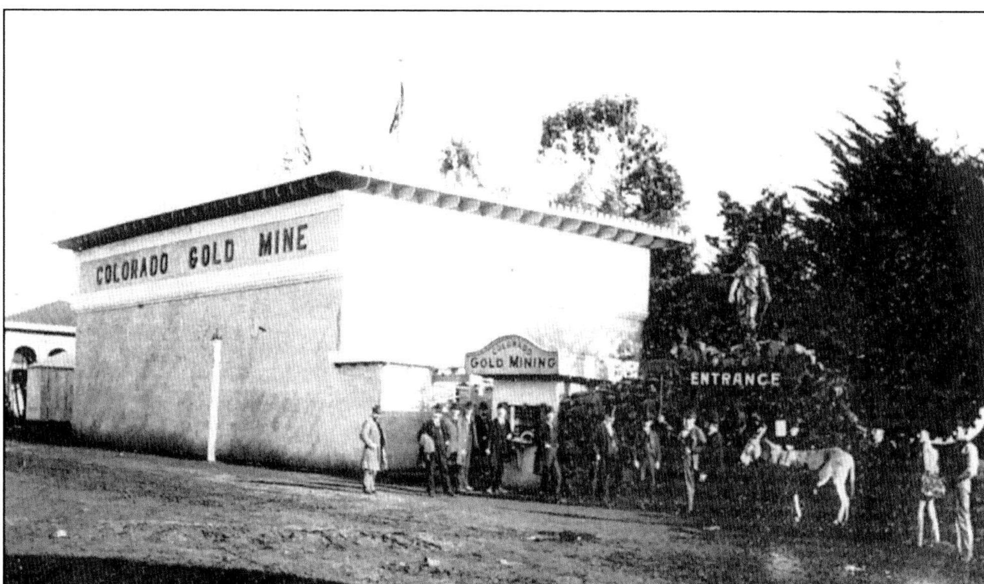

The Colorado Gold Mine was meant to be educational and exciting at the same time, one of only a few instructional exhibits that charged admission. Visitors entered through a realistic shaft entrance to view the workings of a modern mine in full operation, but on a miniature scale.

San Francisco became a metropolis because of mining, but few residents knew exactly how excavating, quarrying, blasting, tunneling, and boring worked. Although the gold now was in the earth, not on it, some hardy folk still panned for it. Four years after the Fair, their attention turned north to Alaska, the last North American gold rush.

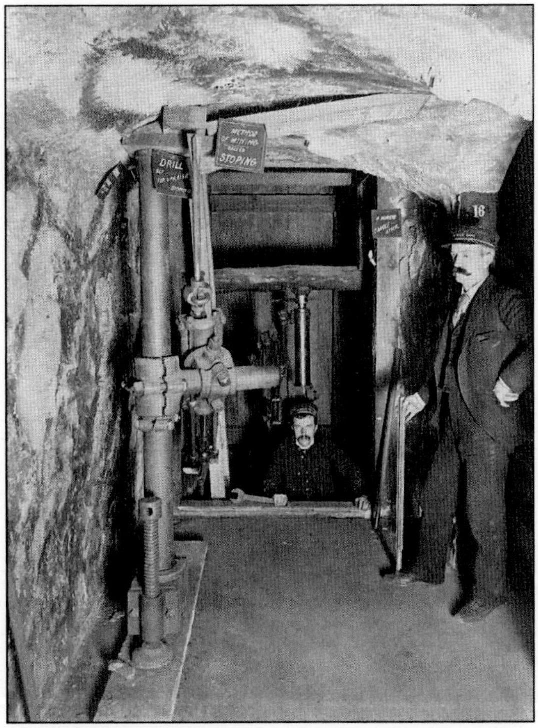

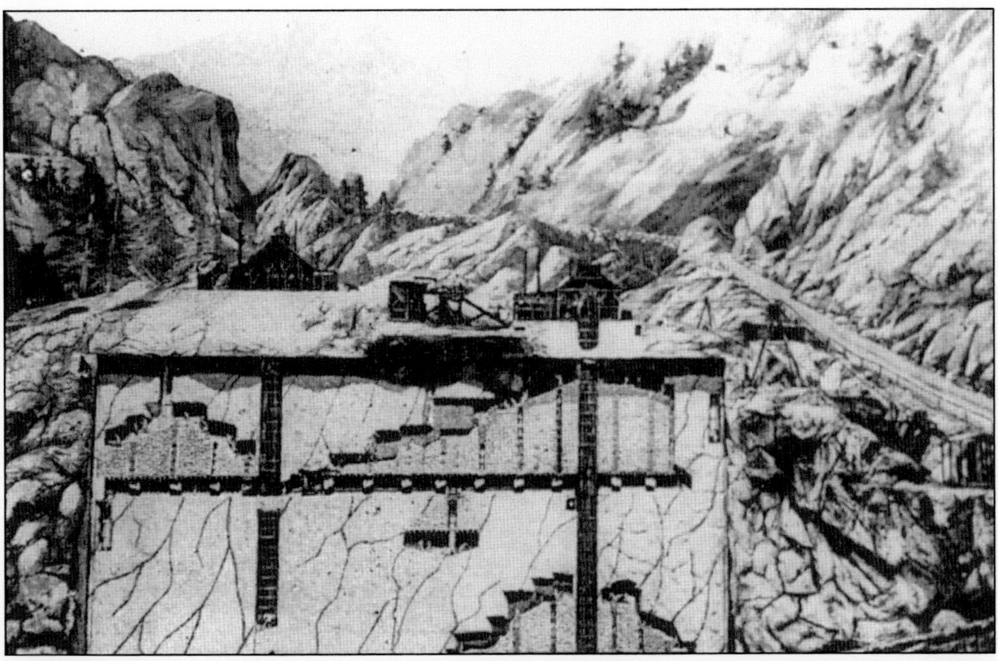

Inside the gold mine, tiny puppets, an inch or so in height, toiled and burrowed in their tunnels and levels to demonstrate the way people, hundreds of feet beneath the surface, searched for gold.

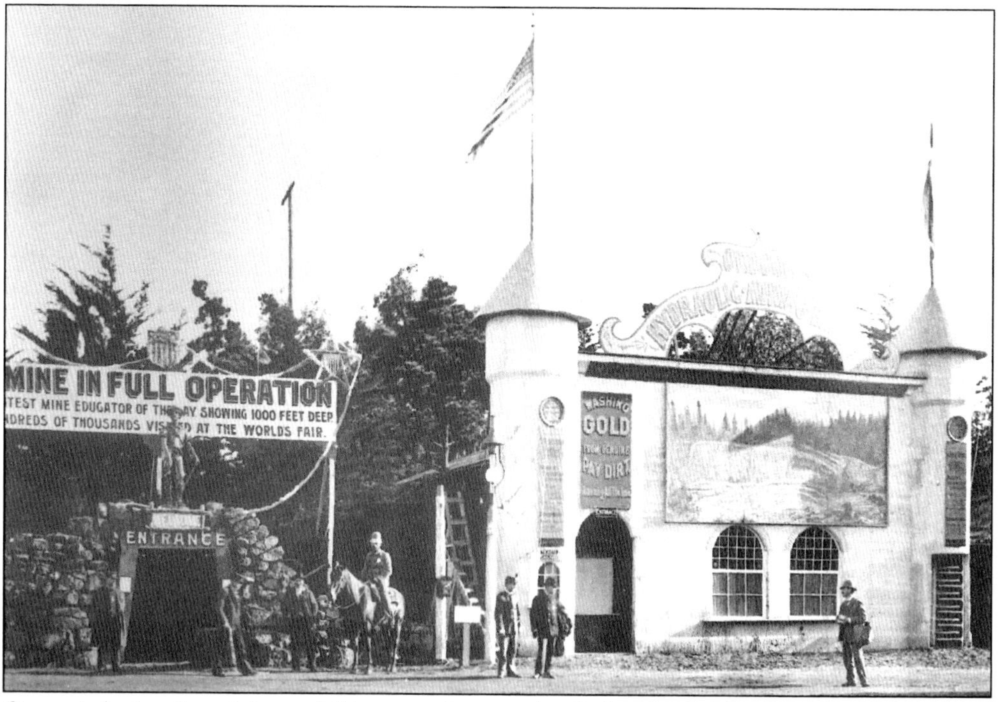

Oregon's hydraulic mining exhibit, next door to the Colorado Gold Mine, showed how this process extracted ore from mountains by washing the mountains away. In California the technique was made illegal in the Sierra Nevadas in 1884, although it persisted in the Klamath-Trinity Range until the 1950s. The water stream from the nozzle could cut down anyone who got in its way.

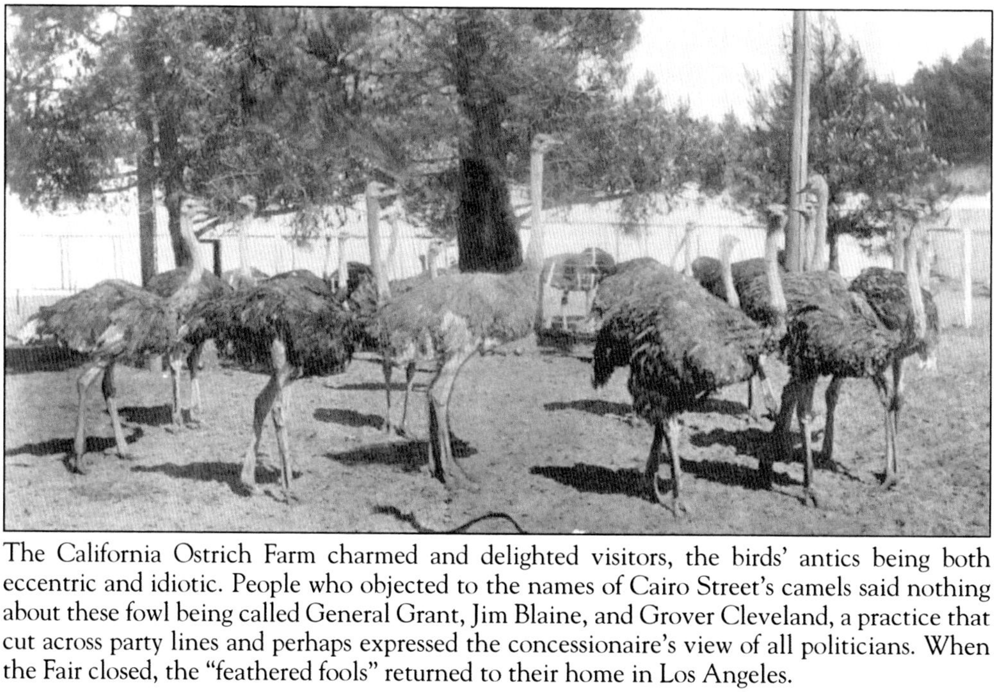

The California Ostrich Farm charmed and delighted visitors, the birds' antics being both eccentric and idiotic. People who objected to the names of Cairo Street's camels said nothing about these fowl being called General Grant, Jim Blaine, and Grover Cleveland, a practice that cut across party lines and perhaps expressed the concessionaire's view of all politicians. When the Fair closed, the "feathered fools" returned to their home in Los Angeles.

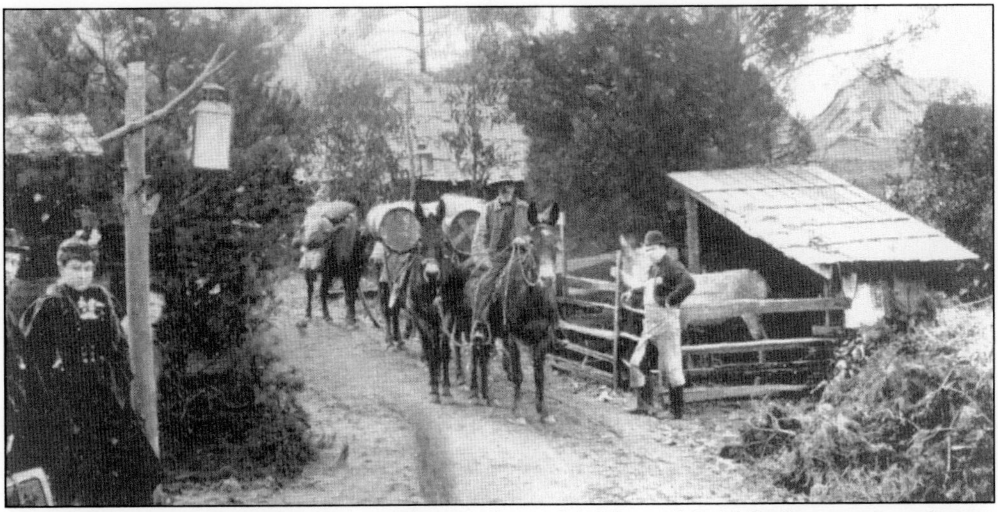

The '49 Mining Camp offered the children and grandchildren of Gold Rush pioneers a glimpse of what life had been like in "the days of old, the days of gold, the days of '49." Walking along dusty streets, they could visit the rowdy enterprises, prospect for gold in the backcountry of the fairgrounds, and, from time to time, witness nostalgic recreations of claim jumping, pistol arbitration, and lynching bees.

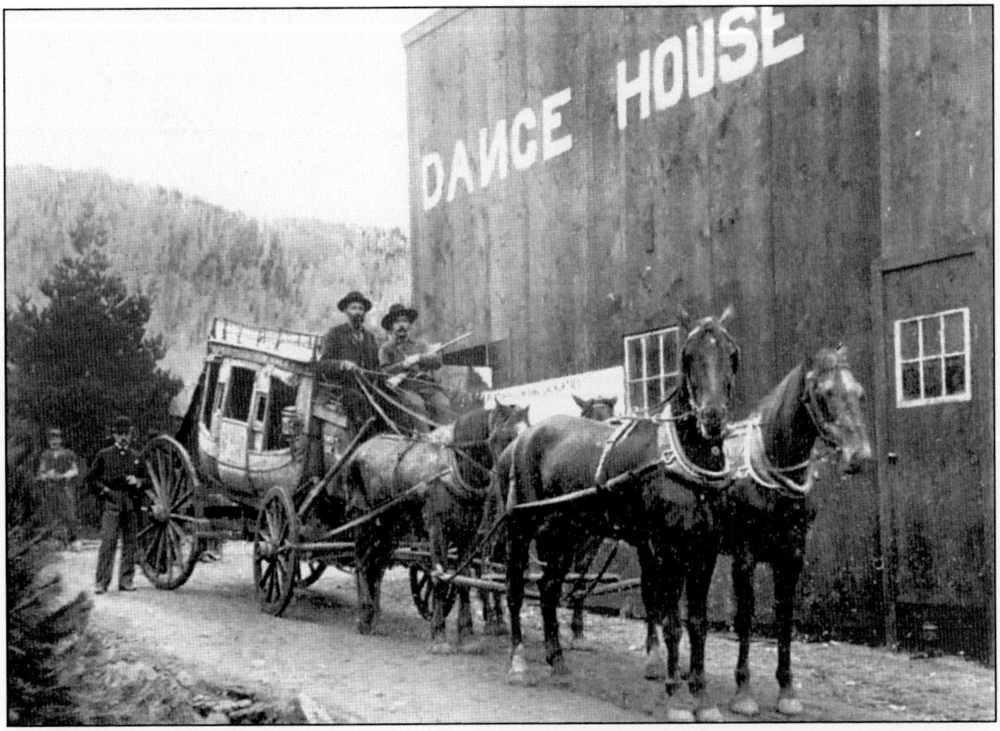

To get in the spirit, some visitors rode to the camp in a stagecoach that made the trip about once an hour from the Administration Building. Years earlier, famed teamster Hank Monk had used it to transport Horace "Go West, young man" Greeley during Greeley's own western travels. Greeley told Monk to slow down so often that his constant reply, "Keep your seat, Horace," became a catchphrase of the day.

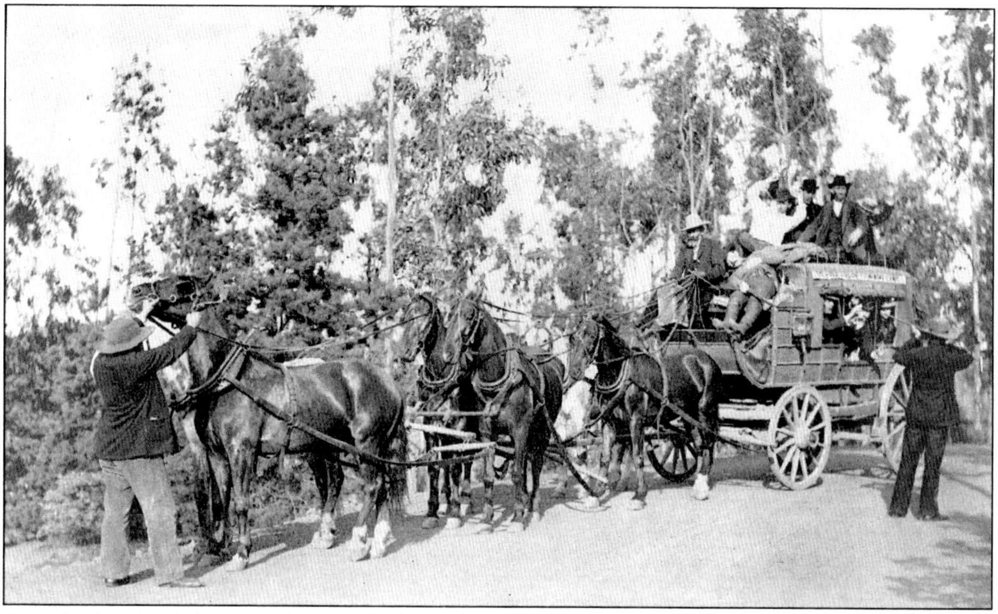

"Bandits" robbed the stage frequently. Here, having shot the guard, they are about to receive the strongbox. Such adventure on a warm winter afternoon, no doubt, was the envy of every passerby. The Fair's only serious accident occurred when the coach overturned, injuring 14 passengers, none seriously. Because stage lines connected many smaller communities in California until World War I, some found less romance in the trek than others.

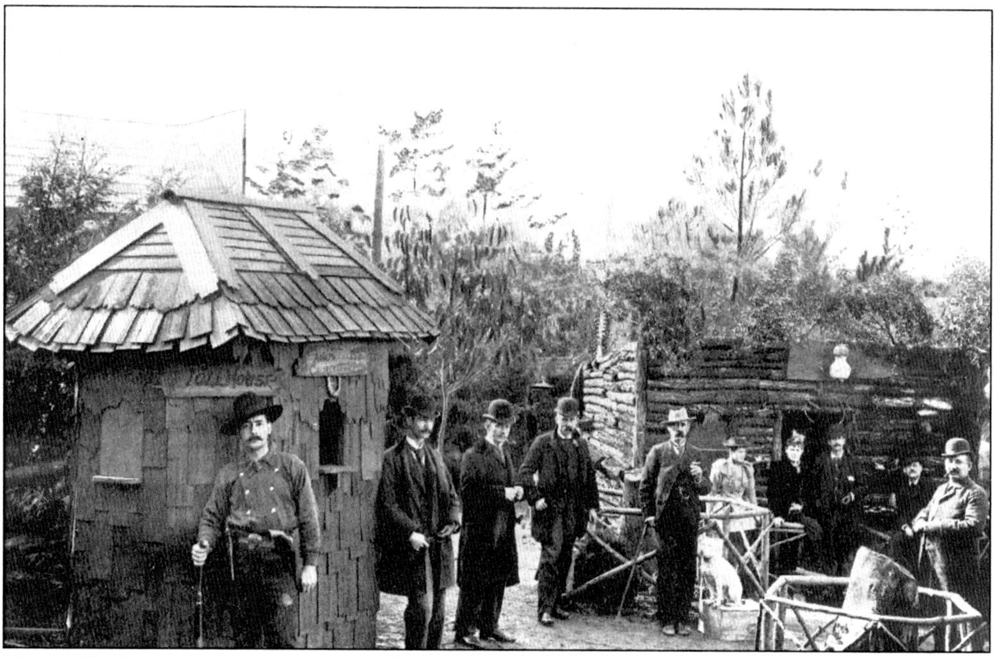

The Toll House at the entrance to the camp had been brought to the Fair from Robinson's Ferry, an early mining community in Calaveras County. Here the "Midwinter Pioneers" could chat with James S. Brown and Israel Evans, among the last living witnesses to the first gold mining in California by other than Native Americans.

After paying the toll of 25¢, the '49ers of '94 could visit the town's livery stable and corral; the Rest for the Weary Hotel; the newspaper office; the saloon and dance hall, which also housed the mayor's offices, the justice of the peace, and the post office, shown here; some original cabins of the '49ers, and some original '49ers, too.

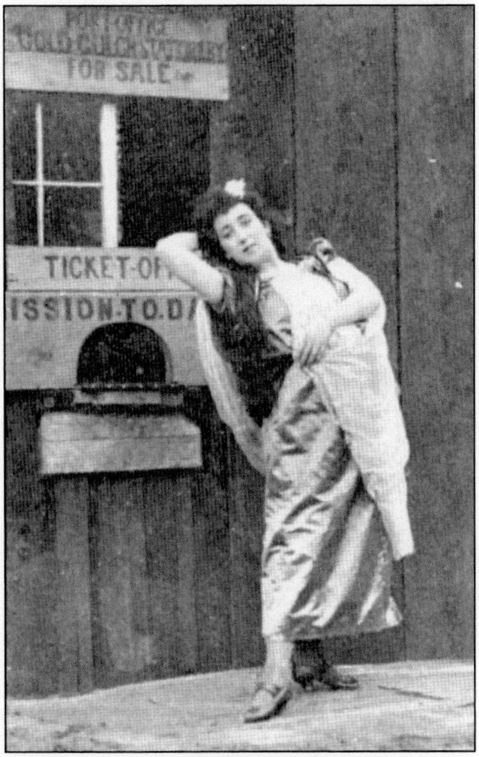

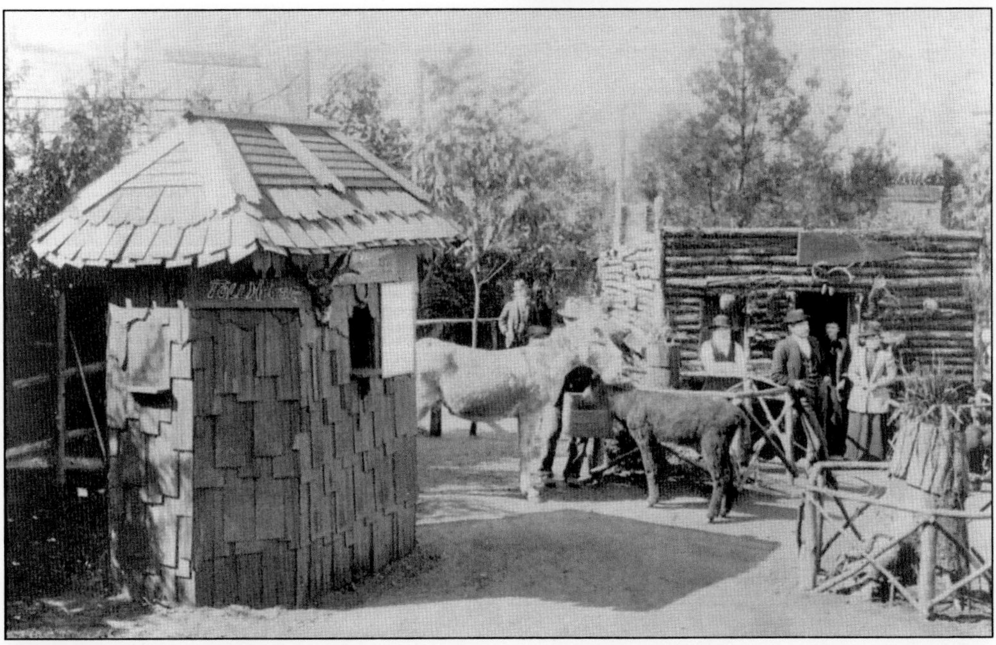

The burros of the '49 Camp gave it an authentic air in more ways than one. Miners referred to them as Sierra-Nevada nightingales and Calaveras canaries, both time honored monikers. The pack animals often followed the stage through the fairgrounds, where, out for a bit of a roam, they might encounter donkeys from Cairo Street, mules from the Hawaiian Village, or a jackass or two from hither and yon.

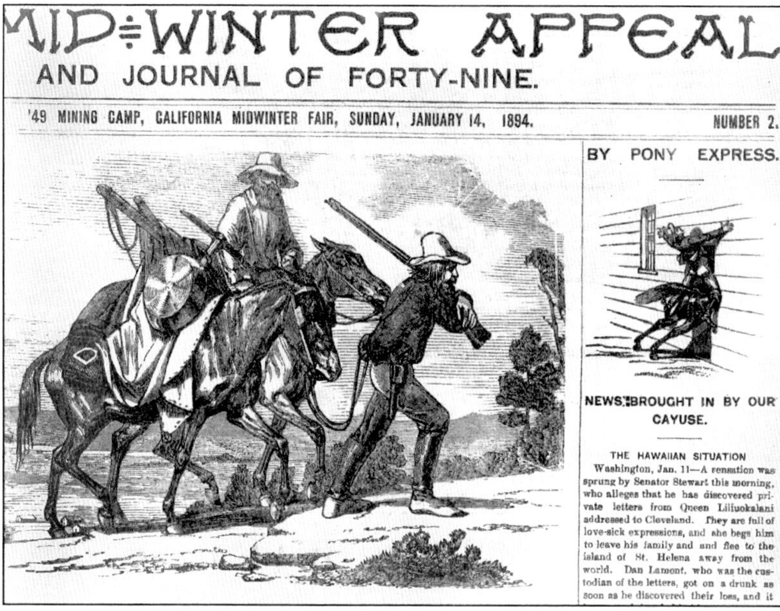

MID-WINTER APPEAL
AND JOURNAL OF FORTY-NINE.

'49 MINING CAMP, CALIFORNIA MIDWINTER FAIR, SUNDAY, JANUARY 14, 1894. NUMBER 2.

BY PONY EXPRESS.

NEWS BROUGHT IN BY OUR CAYUSE.

THE HAWAIIAN SITUATION

Washington, Jan. 11—A sensation was sprung by Senator Stewart this morning, who alleges that he has discovered private letters from Queen Liliuokalani addressed to Cleveland. They are full of love-sick expressions, and she begs him to leave his family and and flee to the island of St. Helena away from the world. Dan Lamont, who was the custodian of the letters, got on a drunk as soon as he discovered their loss, and it

The newspaper office in fact published a weekly newspaper, the *Midwinter Appeal.* Here was news about the Society for the Suppression of Other People's Vices and John L. Sullivan's second championship defeat, "knocked out this time by his wife." San Francisco's own James Corbett, who supplied the first, did better in 1894, successfully defending his title against Charley Mitchell.

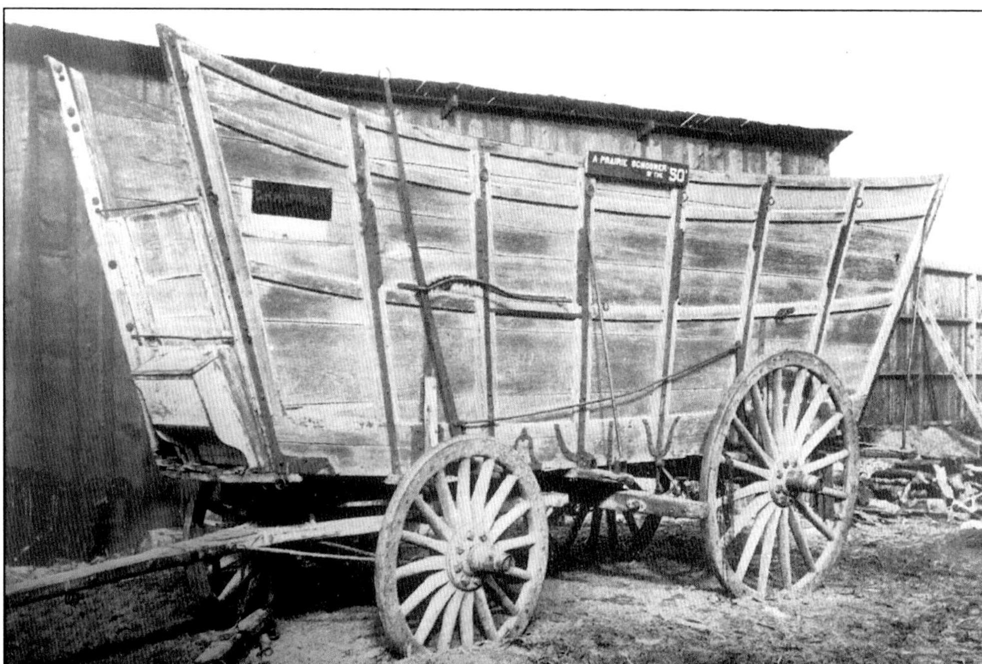

The prairie schooner of the 1850s, an authentic relic of Gold Rush days, was built for ceaseless, stalwart service. Could it have used its tongue to speak, its stories would have been of both grand and hapless adventures on pioneer trails. Overland freighting was a capital intensive business, but also immensely profitable, until the transcontinental railroad ended such enterprises.

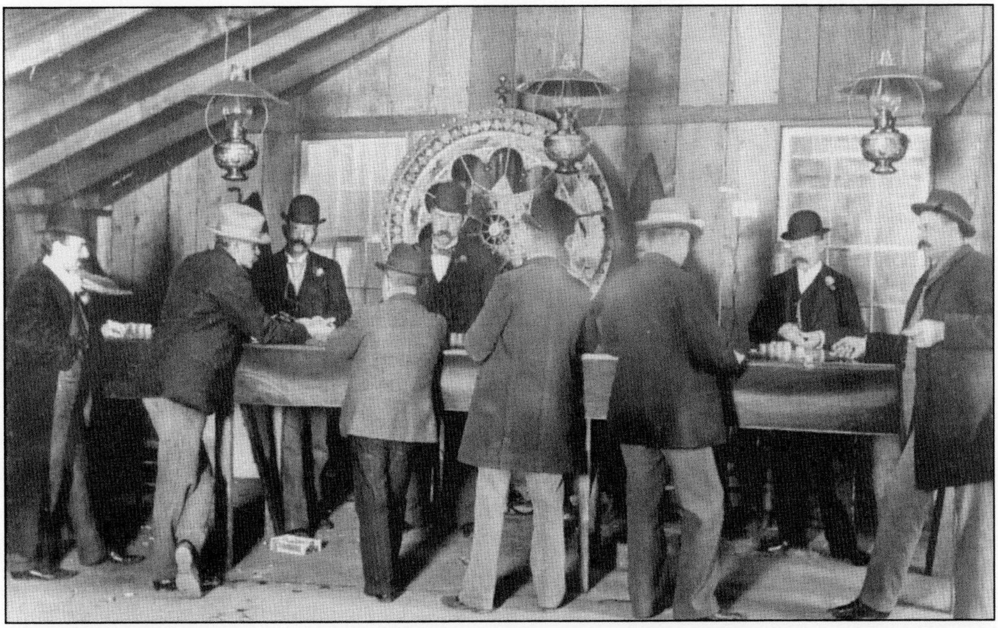

The Gambling House not only featured authentic gaming equipment, it also featured authentic gambling; the Fair's managers decided that otherwise the camp would lack a flavor of genuineness. Patrons regarded the roulette wheel with respect because it was their authentic money on the line.

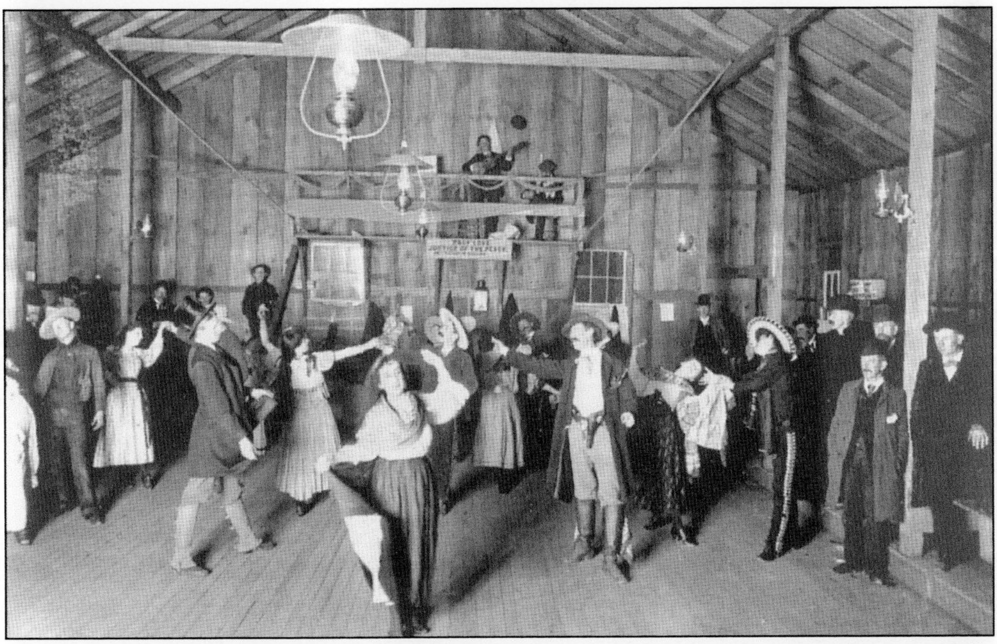

No self-respecting mining town would have a saloon without dancing girls. Here, carefully staged for the photographer, some of establishment's "40 sweet-faced Senoritas" and their admirers perform a stately fandango, made somewhat less regal by its long journey from Spain to the mountains of the mother lode. These Gold Rush gold diggers provided companionship of only the highest moral character.

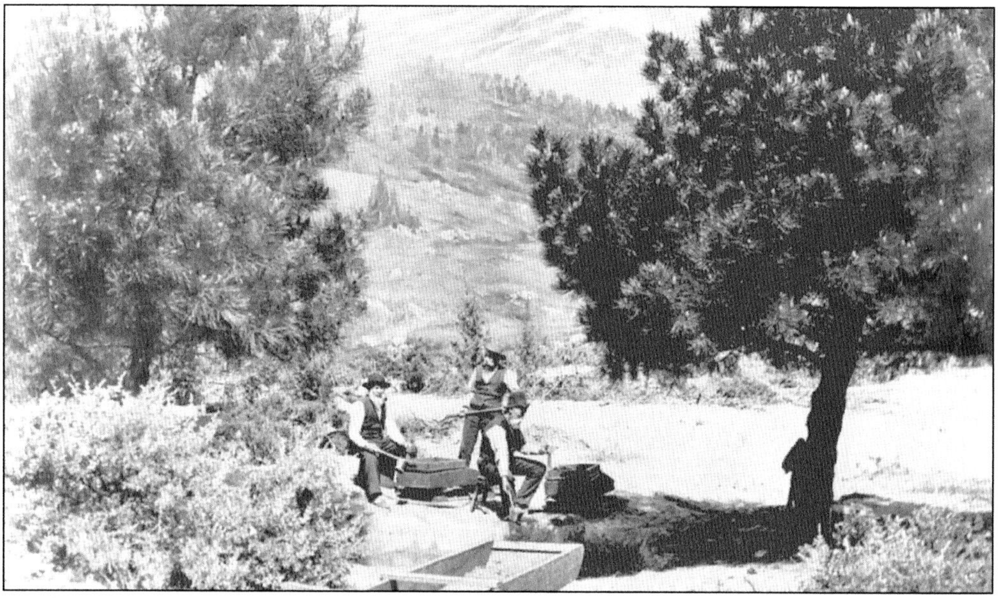

At the far end of the old-time mining camp, cabins, makeshift shanties, and dugouts ran alongside a gulch to where '49ers old and young either panned for gold or used well-worn rockers and long-toms. No gold had yet been found in the park, so their success wasn't assured. Mt. Shasta, in the distance, was a painted canvas backdrop.

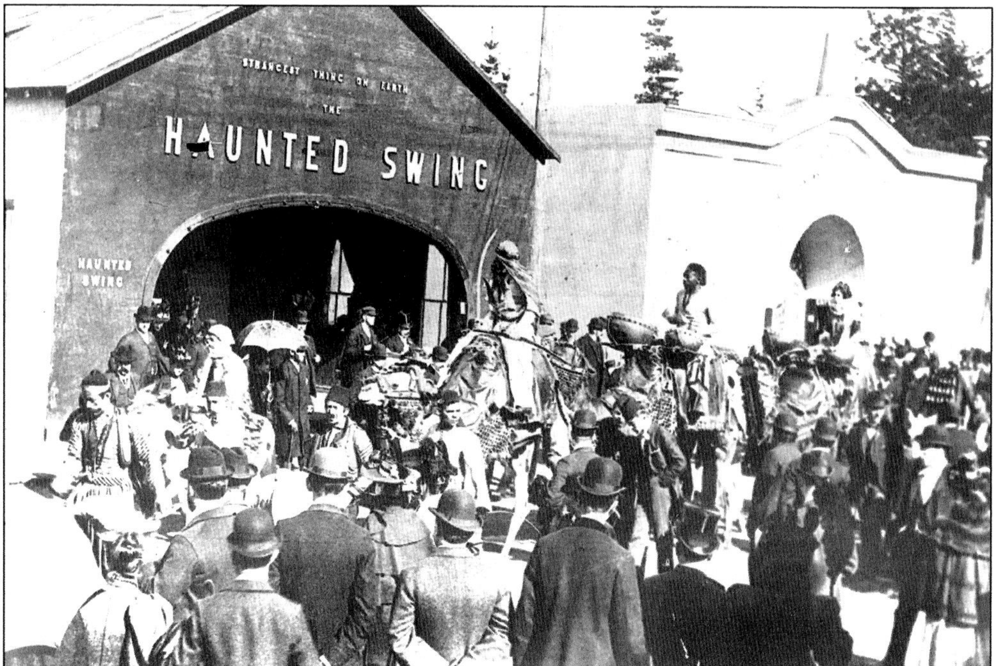

Music was everywhere at the Fair. Here musicians from Cairo Street perform in front of the Haunted Swing, a thrill ride that gave fairgoers the illusion of being turned upside down in a drawing room. Thinking they now were hanging from the ceiling, many crawled under their seats to avoid falling out of them, a precaution that somehow defied logic.

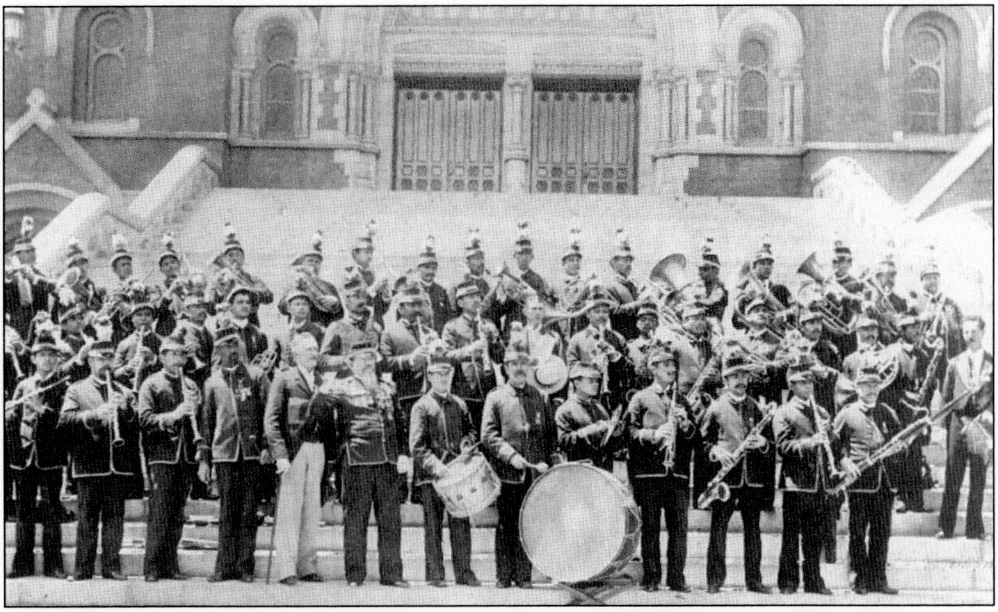

The 8th Regiment Band of the Mexican Army visited the Fair in June. Led by Captain E.B. Payen (first row, sixth from left) the band played an afternoon and an evening concert each day until the Exposition closed officially on July 4. Extremely popular with San Franciscans, the band returned the next year for another series of performances.

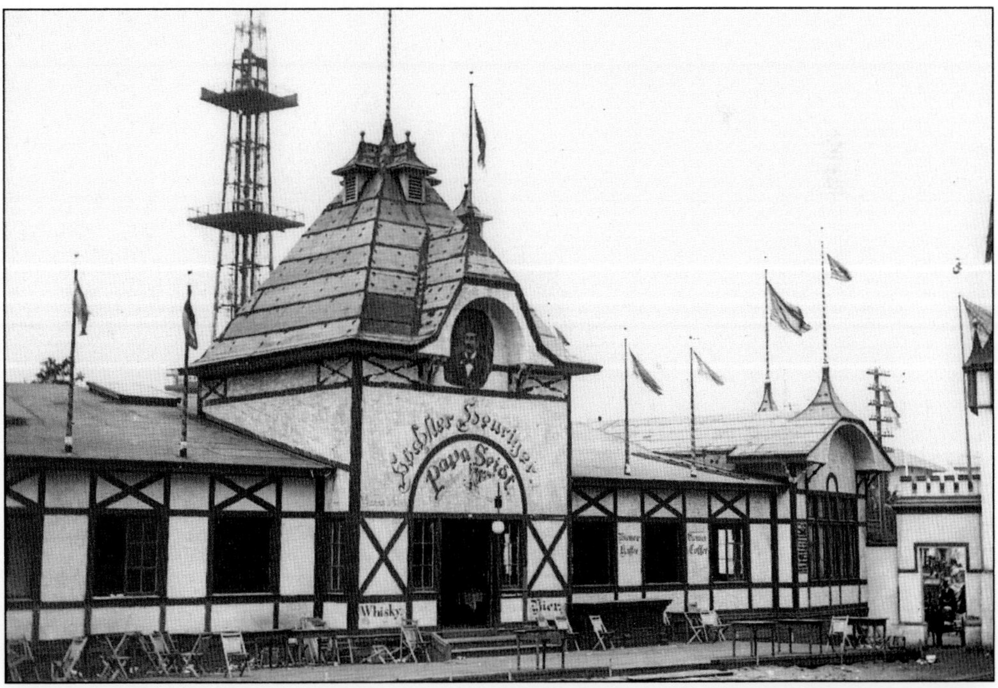

Wonders for the tongue also awaited visitors all over the fairgrounds. There were restaurants in the Chinese, Japanese, and Oriental Villages; in Old Heidelberg; at the Firth Wheel, and the Electric Theater; and at Papa Seidl's in the Vienna Prater, shown here. Patrons also could visit numerous kiosks and vendors, or simply bring a pannier from home and picnic in the park instead.

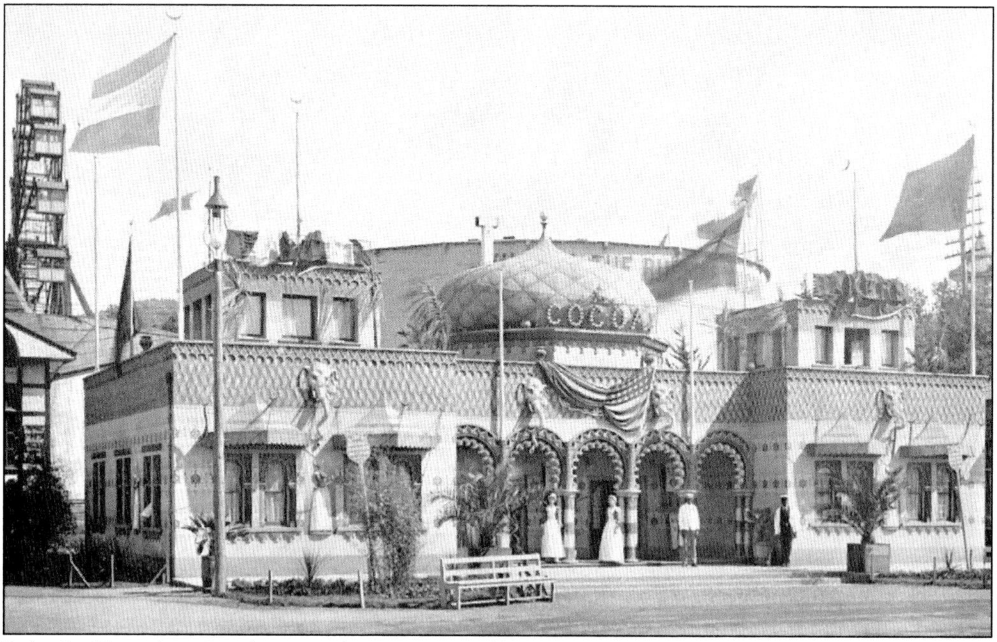

Baker's Chocolate Pavilion, with its mosque-like dome and its facade adorned with life-size elephant heads, evoked the exotic realms that were home to the beloved bean. No surprise that the walls were painted light and dark shades of brown. For reasons unknown, attendants didn't dress to match the decor; they looked more like modern Priscillas than anyone else.

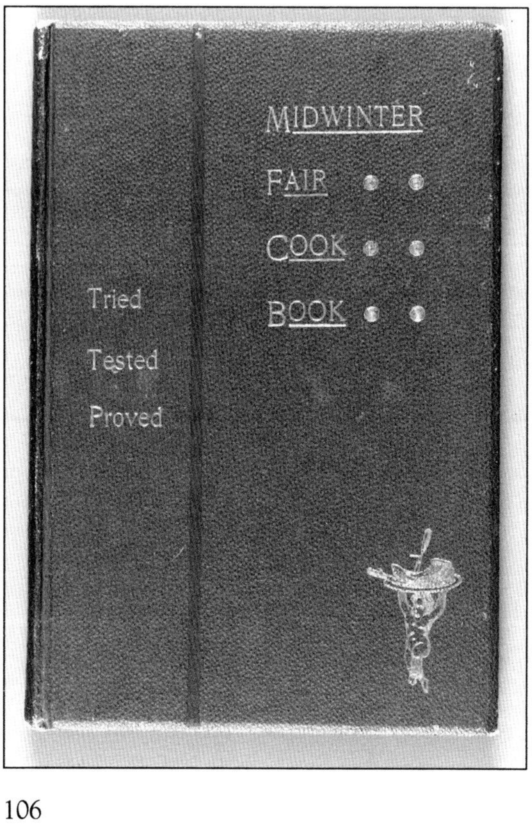

The *Midwinter Fair Cookbook* contained recipes "contributed by ladies of San Francisco and other cities." Typical was their directions for fried squirrel: "Break into pieces after washing well and then dip each separately into flour and fry in *hot* lard." The ladies felt no need to provide temperatures or cooking times. At least they didn't begin, "First, catch a squirrel."

Collecting souvenir spoons was one of the great fads of the 1890s. More than 100 different styles exist for the Midwinter Fair. Popular subjects were buildings, bears, and miners. Caveat emptor: in those less standardized and regulated times, some spoons marked sterling had no silver content at all.

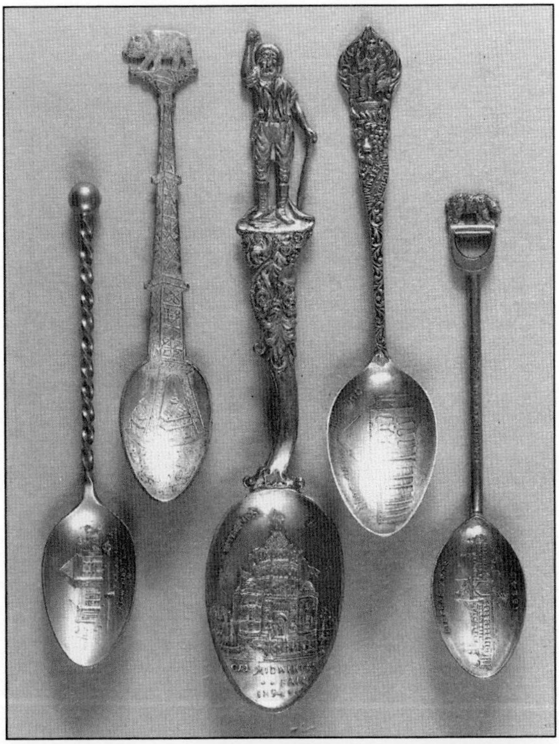

When the Fair closed, Adolph Sutro, San Francisco's mayor-elect, bought the Camera Obscura, the Firth Wheel, Dante's Inferno, the Mirror Maze, and other Midway attractions. He moved them to the Sutro Baths, which he opened in 1896. In addition to its gigantic swimming pool, it housed restaurants, natural history exhibits, art galleries, and an amphitheater. Only the Camera Obscura survives.

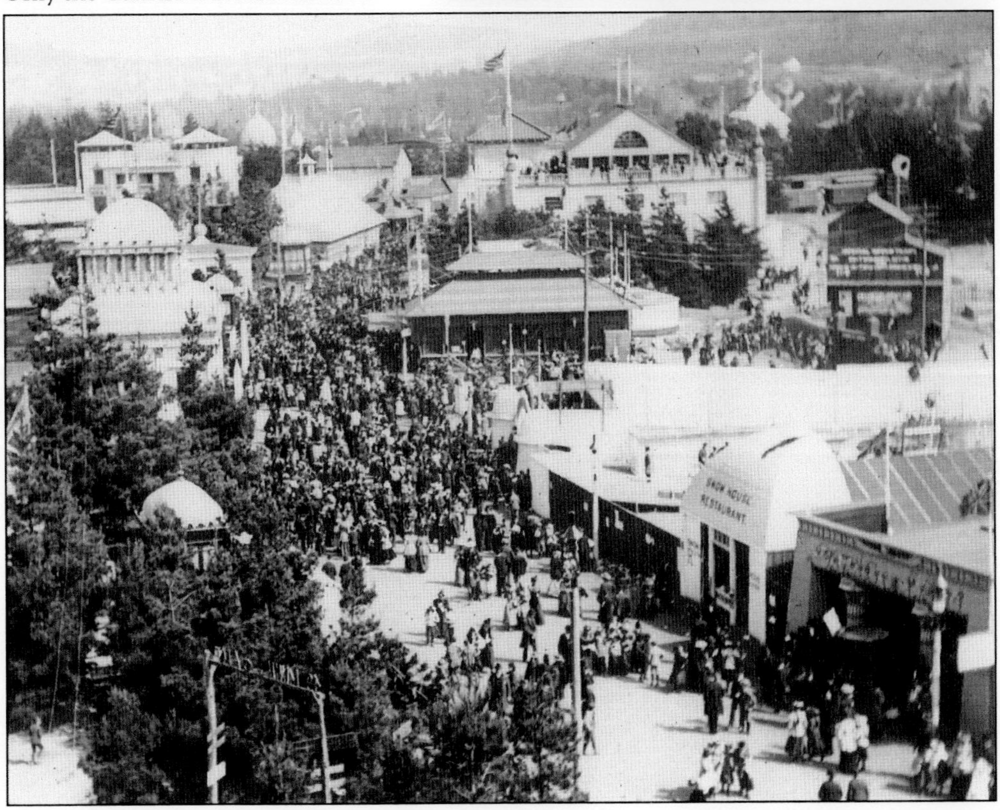

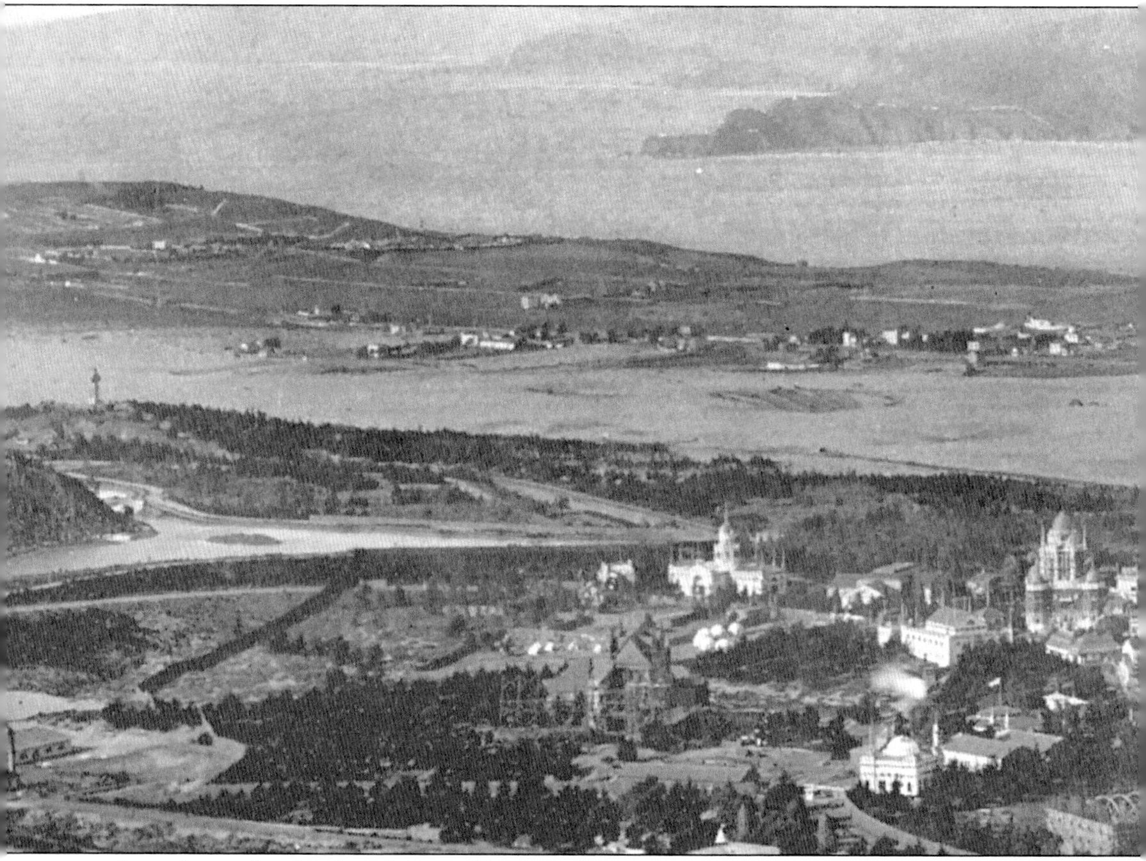

Seen during the day from Mount Olympus, the entire Fair and its surroundings were visible in detail for miles. The dome of the Administration is seen on the left, the Electric Tower rising high above all other structures. To its right, the Egyptian fantasy of the Fine Arts Building seems dwarfed by the Cyclorama of Kilauea across the Court of Honor. At the far right, the massiveness

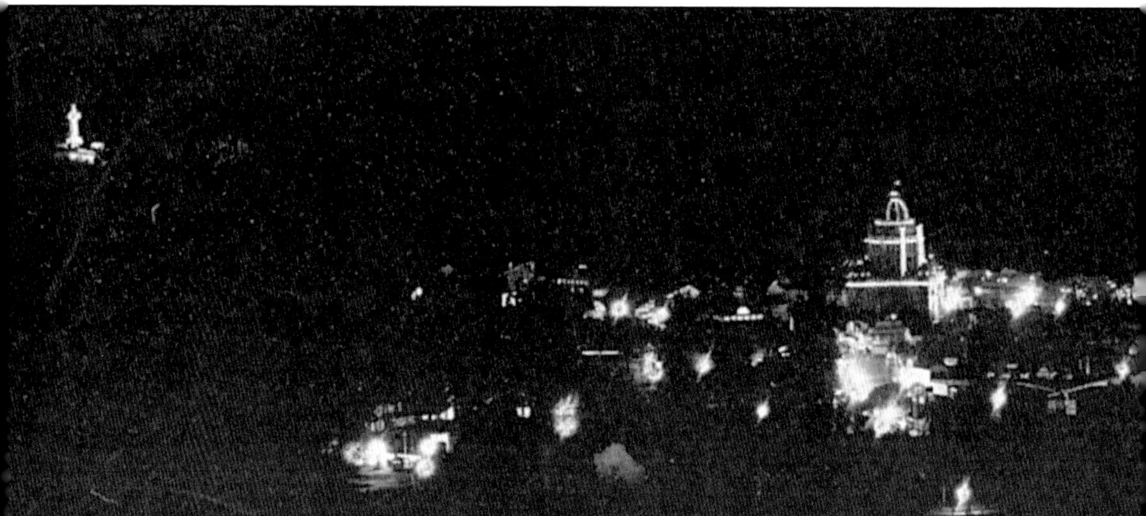

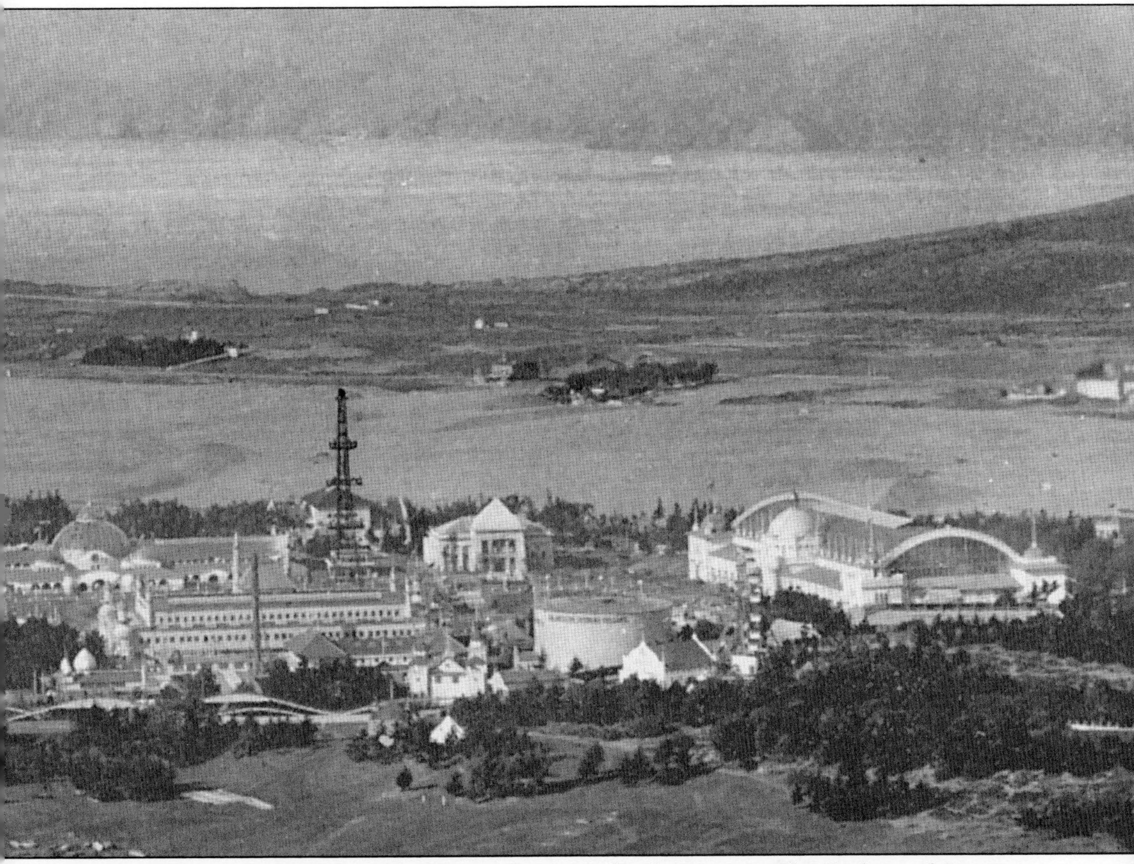

of the Manufactures Building is clearly apparent. In the distance are the Richmond District, the Bay, and the Marin Headlands. At night, however, only the outlines of the buildings were visible, making the Fair appear to be floating in a world of its own.

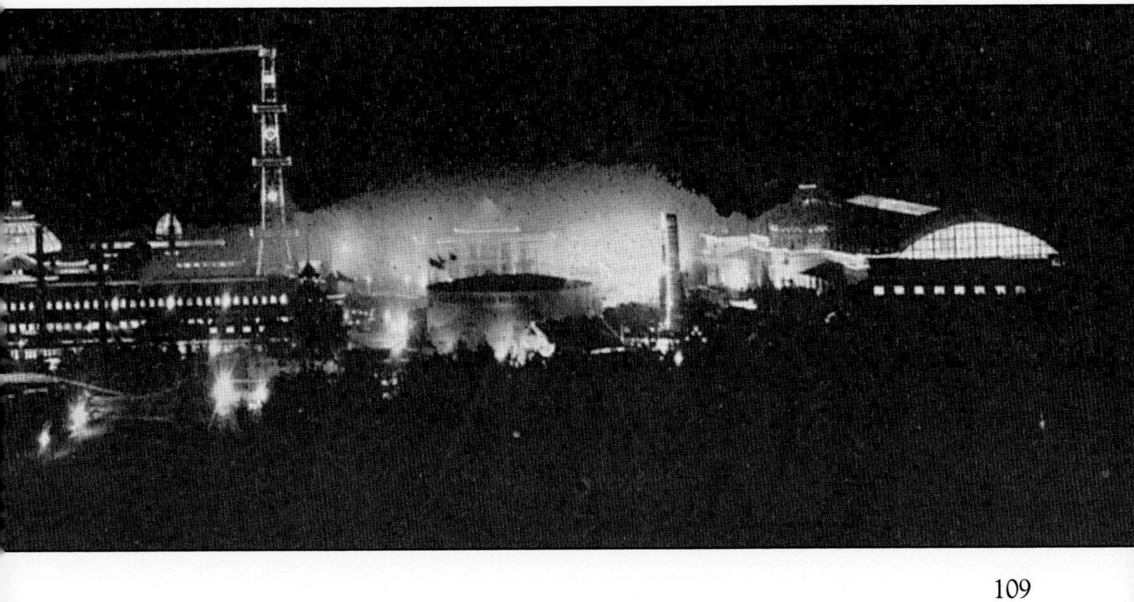

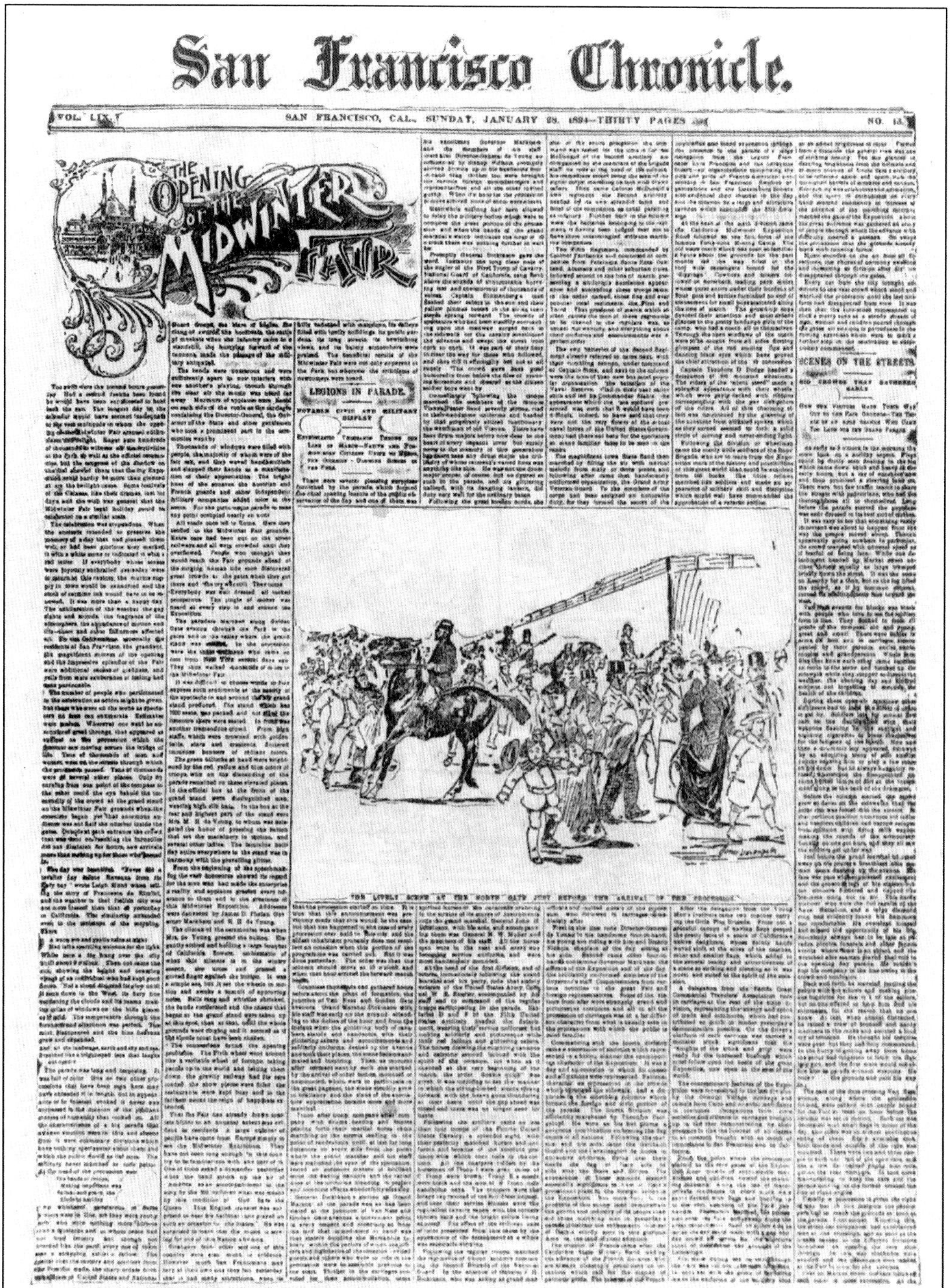

San Francisco Chronicle.

VOL. LIX. SAN FRANCISCO, CAL., SUNDAY, JANUARY 28, 1894—THIRTY PAGES NO. 13.

THE OPENING OF THE MIDWINTER FAIR

LEGIONS IN PARADE.

SCENES ON THE STREETS.

Special days at the Fair required special mementos. Children and adults who visited the Exposition on Chronicle Day received a miniature newspaper as a souvenir. It reproduced the front page of the first *Chronicle*, the front page of the paper's Midwinter Exposition edition, and two pages that showed the Fair's principle buildings, concessions, and features. The newspaper distributed a similar keepsake at the Panama-Pacific International Exposition in 1915.

Five

Fetes, Congresses, Festivals, and Farewells

Whatever their stated ideals and lofty aims, the managers of America's expositions realized that, to be successful, they had to get people past the turnstiles, preferably more than once. Beginning with the Centennial Exhibition of 1876, one favorite and effective plan was to host special fete and festival days that celebrated specific nations, states, counties, cities, professions, trades, organizations, or anniversaries—whatever would attract enough new and returning visitors to make it worthwhile. Another method was to convene congresses to discuss the important issues of the day, which also added prestige and importance to the goings on. The Midwinter Fair used both. Almost every day honored one group or another; many days honored more than one. The Fair also hosted meetings, conferences, and conventions. The two approaches have continued at world's fairs to the present.

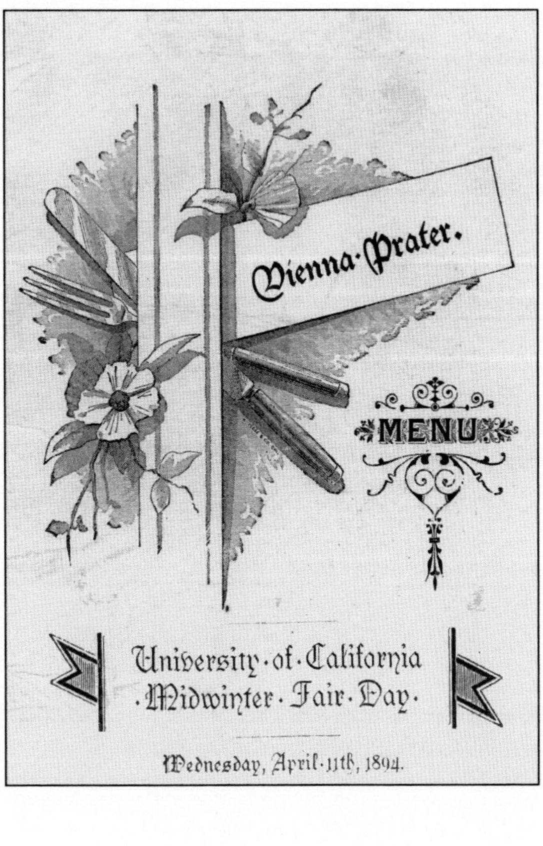

Special days required special invitations, programs, ribbons, badges, and other keepsakes. Fairgoers who attended University of California Day, April 11, received this menu as a souvenir of their evening's banquet. They supped on Consommé Royal, Baked Bass a l'Italienne, potatoes, Sirloin of Beef a la Chambord, string beans, roast chicken, cabinet pudding, fruit, and cheese. Those still awake then heard music by Beethoven, Mascagni, Schubert, Sullivan, Conradi, Zaber, and Weber.

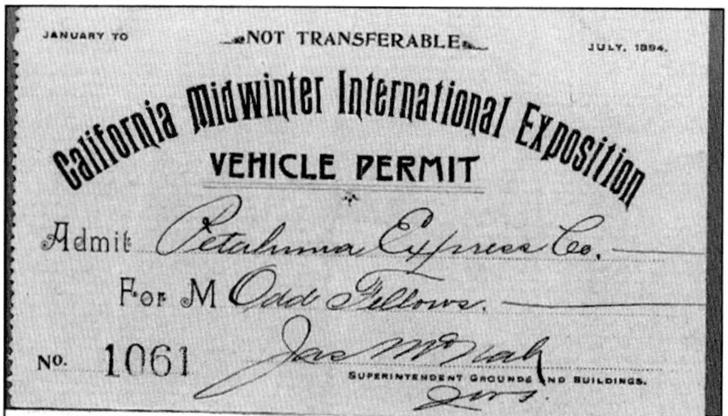

JANUARY TO NOT TRANSFERABLE JULY, 1894.

California Midwinter International Exposition

VEHICLE PERMIT

Admit Petaluma Express Co.

For M Odd Fellows.

No. 1061

Superintendent Grounds and Buildings.

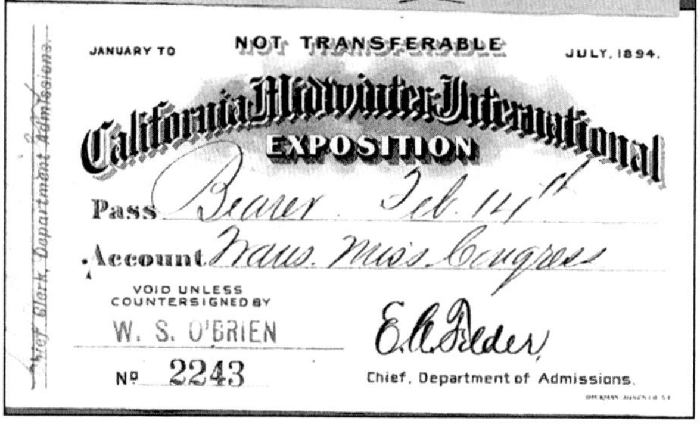

JANUARY TO NOT TRANSFERABLE JULY, 1894.

California Midwinter International Exposition

Pass Bearer Feb. 14th

Account Trans. Miss. Congress

VOID UNLESS COUNTERSIGNED BY

W. S. O'BRIEN

No 2243

Chief, Department of Admissions.

Chief Clerk, Department Admissions.

Special days also required bunting, banners, signs, and special attention to detail. The vehicle permit here allowed the Petaluma Express Company onto the grounds to deliver necessary goods for the Odd Fellows Celebration at the Fair during the golden anniversary encampment. The bearer pass enabled a participant of the Trans-Mississippi Congress to visit the Exposition as one of a larger group of delegates.

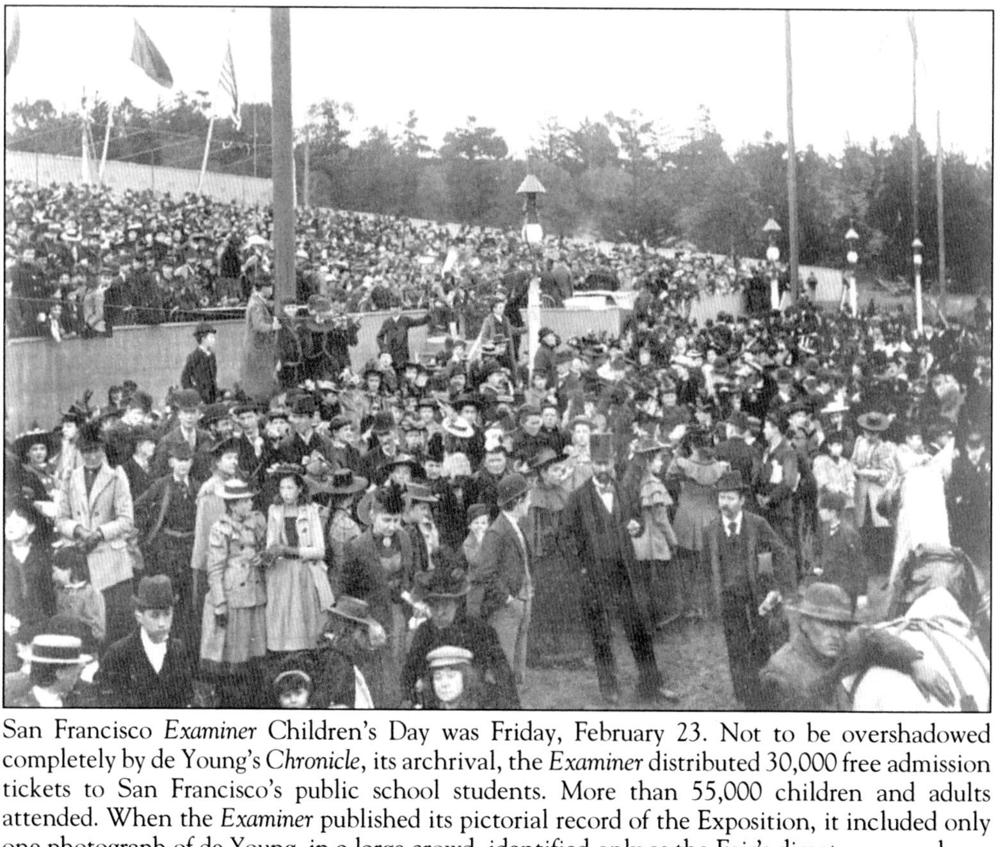

San Francisco *Examiner* Children's Day was Friday, February 23. Not to be overshadowed completely by de Young's *Chronicle*, its archrival, the *Examiner* distributed 30,000 free admission tickets to San Francisco's public school students. More than 55,000 children and adults attended. When the *Examiner* published its pictorial record of the Exposition, it included only one photograph of de Young, in a large crowd, identified only as the Fair's director-general.

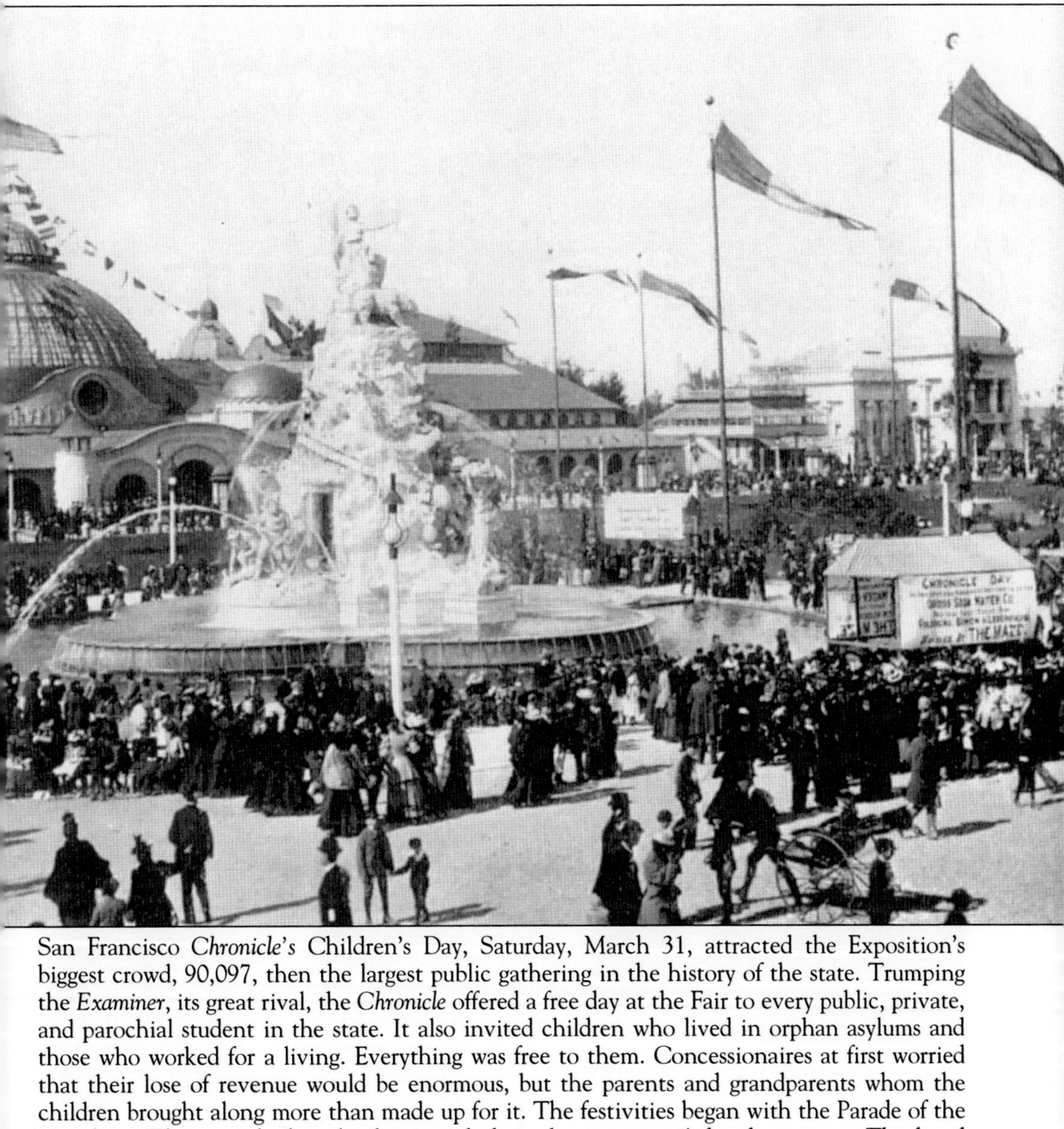

San Francisco *Chronicle's* Children's Day, Saturday, March 31, attracted the Exposition's biggest crowd, 90,097, then the largest public gathering in the history of the state. Trumping the *Examiner*, its great rival, the *Chronicle* offered a free day at the Fair to every public, private, and parochial student in the state. It also invited children who lived in orphan asylums and those who worked for a living. Everything was free to them. Concessionaires at first worried that their lose of revenue would be enormous, but the parents and grandparents whom the children brought along more than made up for it. The festivities began with the Parade of the Newsboys. They marched to the fairgrounds from the newspaper's headquarters at Third and Market Streets in downtown San Francisco.

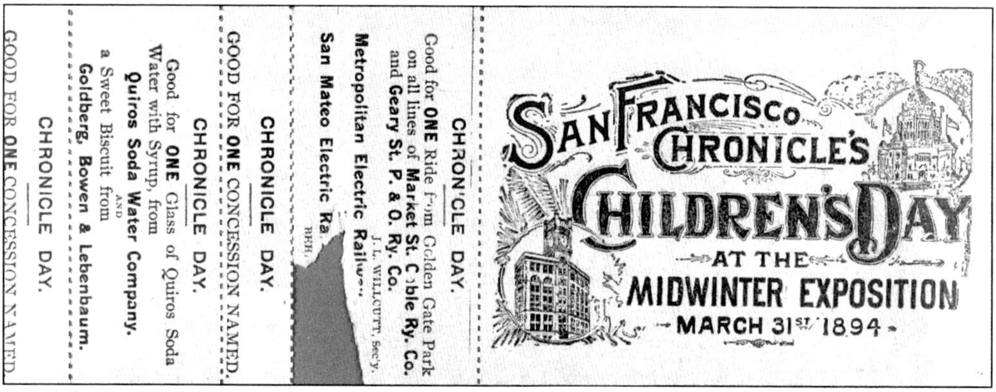

SAN FRANCISCO CHRONICLE'S CHILDREN'S DAY AT THE MIDWINTER EXPOSITION MARCH 31ST 1894.

CHRONICLE DAY.
Good for ONE Ride from Golden Gate Park on all lines of Market St. Cable Ry. Co. and Geary St. P. & O. Ry. Co.
J. L. WILLCUTT, Sec'y.

Metropolitan Electric Railw
San Mateo Electric Ra

CHRONICLE DAY.
GOOD FOR ONE CONCESSION NAMED.

CHRONICLE DAY.
Good for ONE Glass of Quiros Soda Water with Syrup, from Quiros Soda Water Company.
AND
a Sweet Biscuit from Goldberg, Bowen & Lebenbaum.

CHRONICLE DAY.
GOOD FOR ONE CONCESSION NAMED.

The admission ticket for *The Chronicle's* Children's Day had coupons for transportation to and from Golden Gate Park, entry to the fairgrounds, and all the essentials for a perfect day at the Fair. This ticket holder apparently used the free ride coupon to get to the Exposition—or perhaps simply discarded it to use a concession coupon; how he or she got home will remain a mystery.

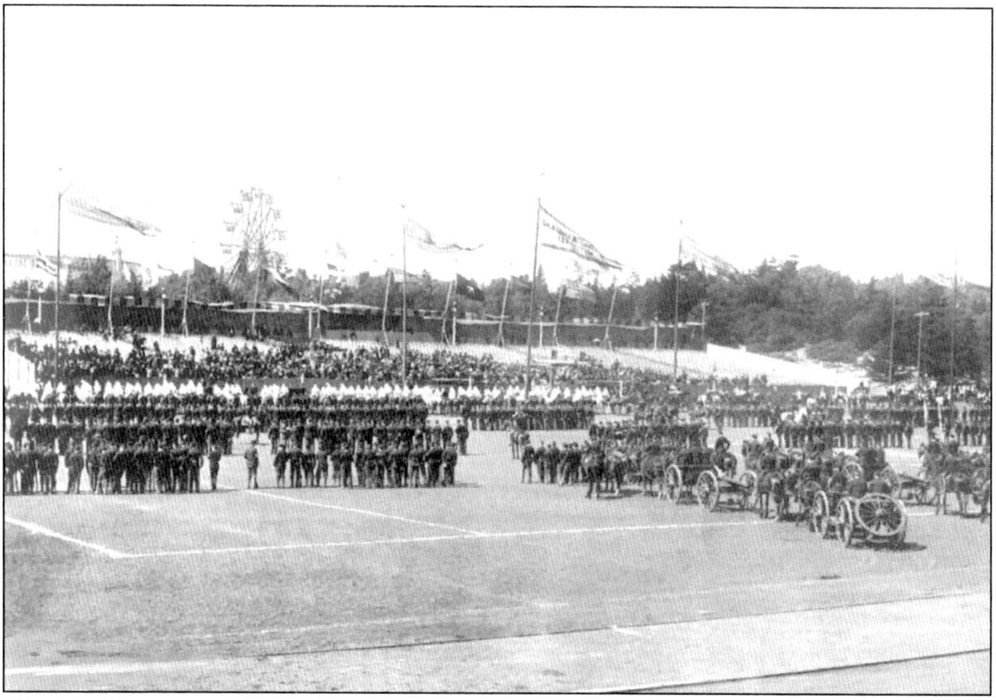

The San Francisco Carnival was April 17–19. The first day featured a parade from the Ferry Building to the fairgrounds. Although the grandstand seems mostly empty, this was due, apparently, to "the thoughtfulness of the Exposition management in providing accommodations for more than took advantage of them," not because very few people showed up on a Tuesday afternoon. The carnival's second day featured a tournament, the third a *Carnival Bal Masque*.

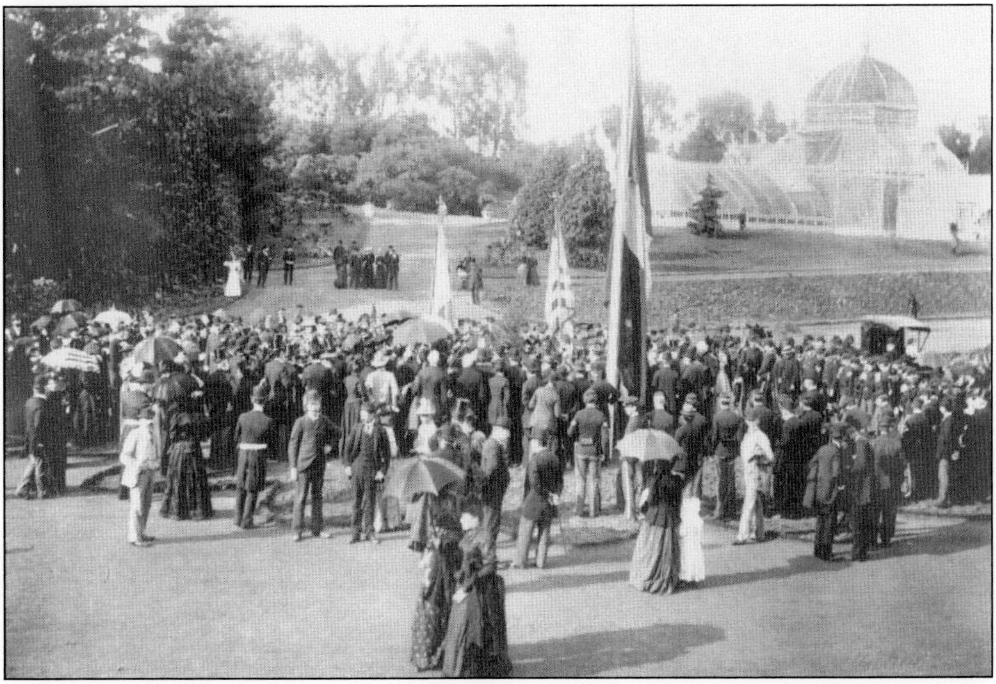

On April 19, the anniversary of the Battles of Lexington and Concord, the Daughters of the American Revolution, who curiously seem absent in the photograph, planted a liberty tree near Golden Gate Park's Conservatory of Flowers. Their sequoia was rooted in soil sent to them from Mt. Vernon, Montecello, Independence Hall, Valley Forge, Lafayette's Tomb, Yorktown, and dozens of other similar sites.

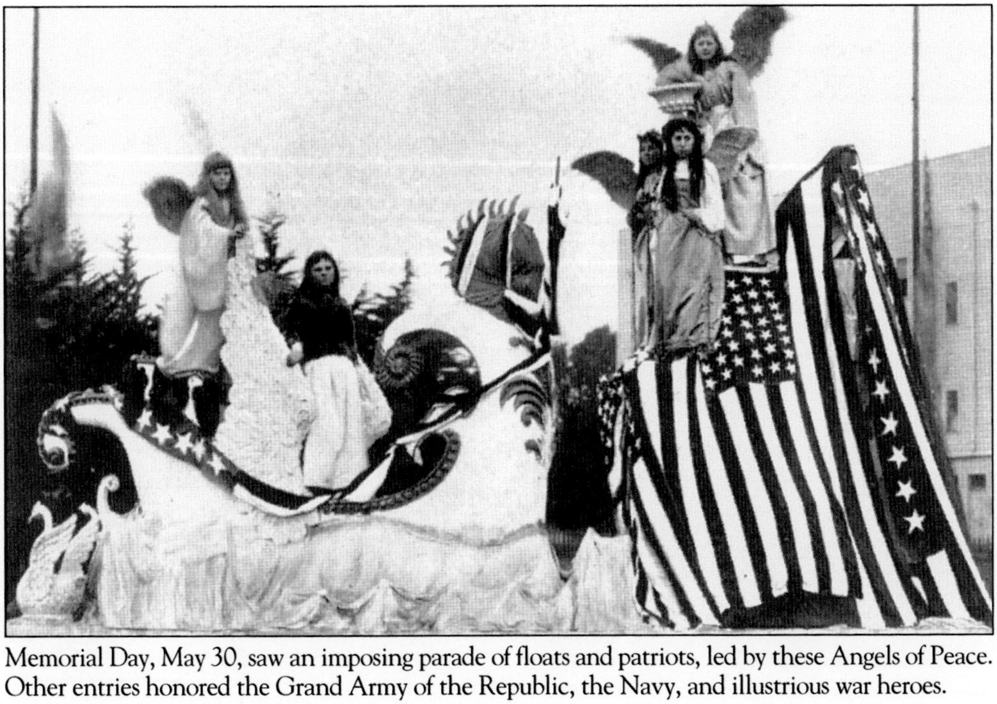

Memorial Day, May 30, saw an imposing parade of floats and patriots, led by these Angels of Peace. Other entries honored the Grand Army of the Republic, the Navy, and illustrious war heroes.

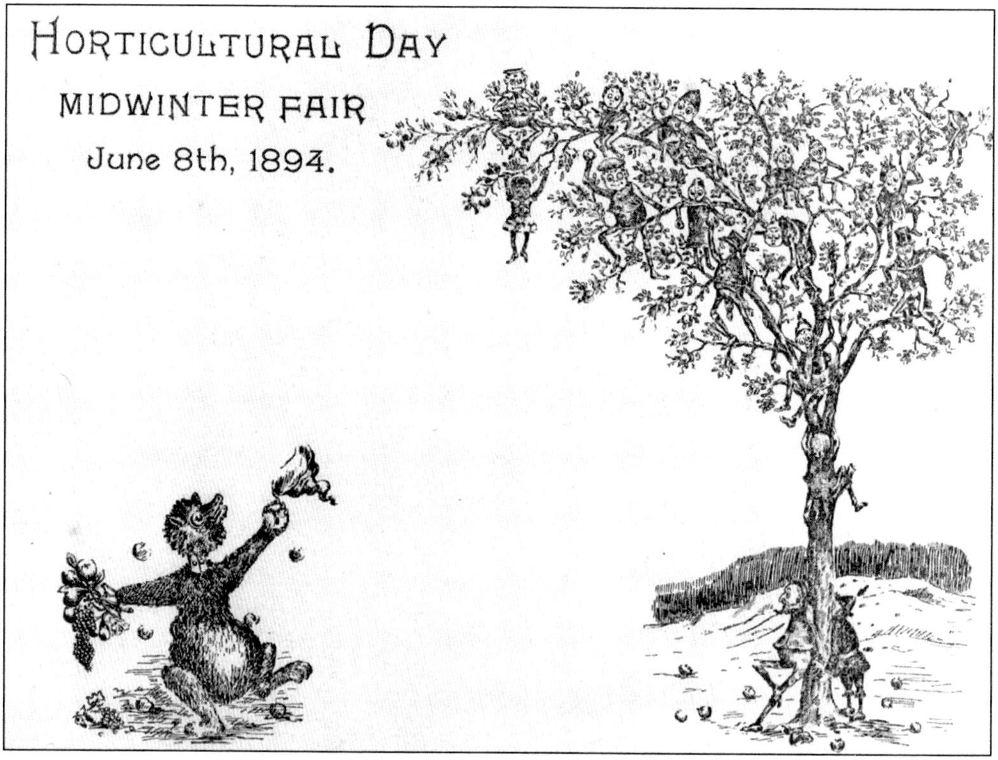

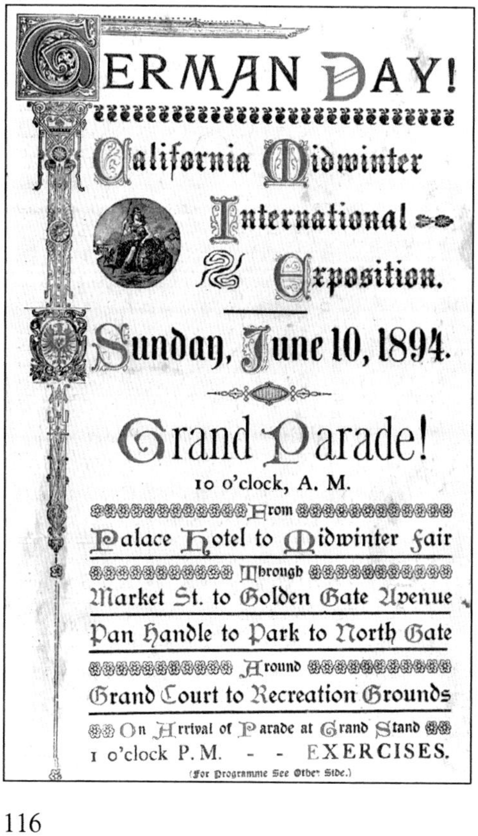

Horticultural Day, June 8, managed to involve Park Superintendent McLaren in the activities of the Fair, when he assisted Miss Nellie Boyd with the planting of a tree. The ceremony was followed inside the Horticulture Building with a program of speeches, songs, and music by Cassasa's Exposition Band, by which time McLaren had left. Palmer Cox, famed for writing and illustrating the beloved *Brownies* stories, created the whimsical invitation.

German Day, Sunday, June 10, began at 10 a.m. with a Grand Parade to the fairgrounds. Exercises at the fairgrounds included music, prayers, addresses, recitations, and gymnastics, followed by continuous concerts at both the Vienna Prater and Old Heidelberg. Fireworks commenced at 8 p.m., the Grand Ball at 9 p.m. Skylarks who stayed until the wee hours found that the electric and cable car lines had extended their hours of service so they could get home. Attendance reached 68,643. The hand bill, left, gave all the particulars.

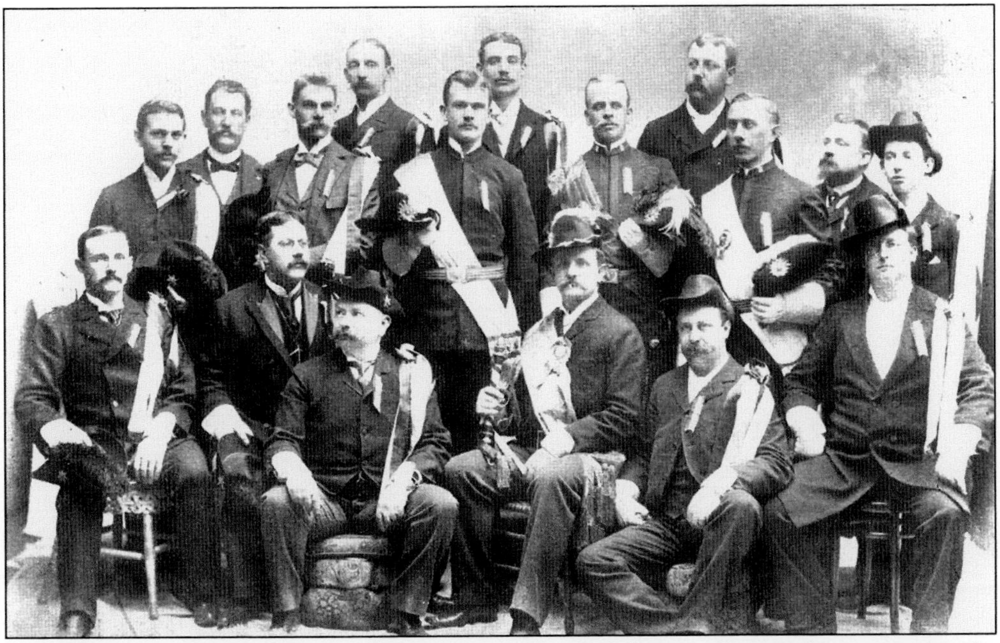

The German Day committee posed for their portrait in full array. During the mauve decade, men were great joiners of organizations. Whether for fun, fraternity, or politics—or just to get out of the house for an evening—many males belonged to one or more professional, ethnic, or social societies. Only those who worried that Papa would return home neither soon nor sober distressed themselves of such camaraderie.

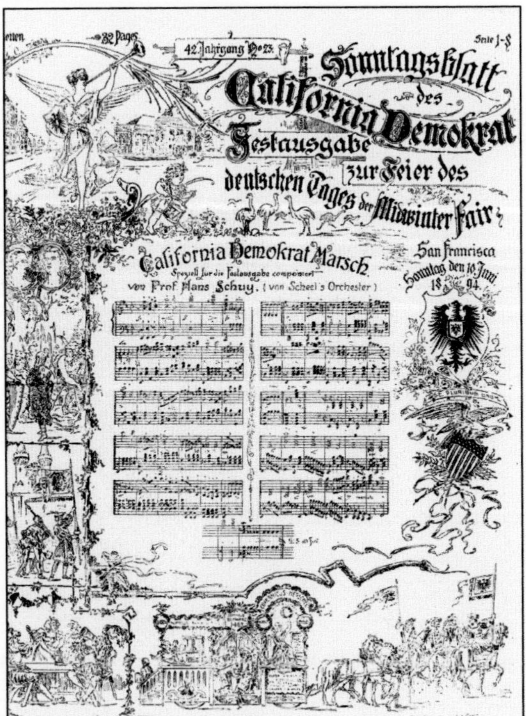

German Day had its own music, as well as the usual souvenir ribbons, badges, pins, programs, and cards. The *California Demokrat*, the leading German language newspaper, published a special edition with the *California Demokrat March*, especially written for the celebration, on the cover.

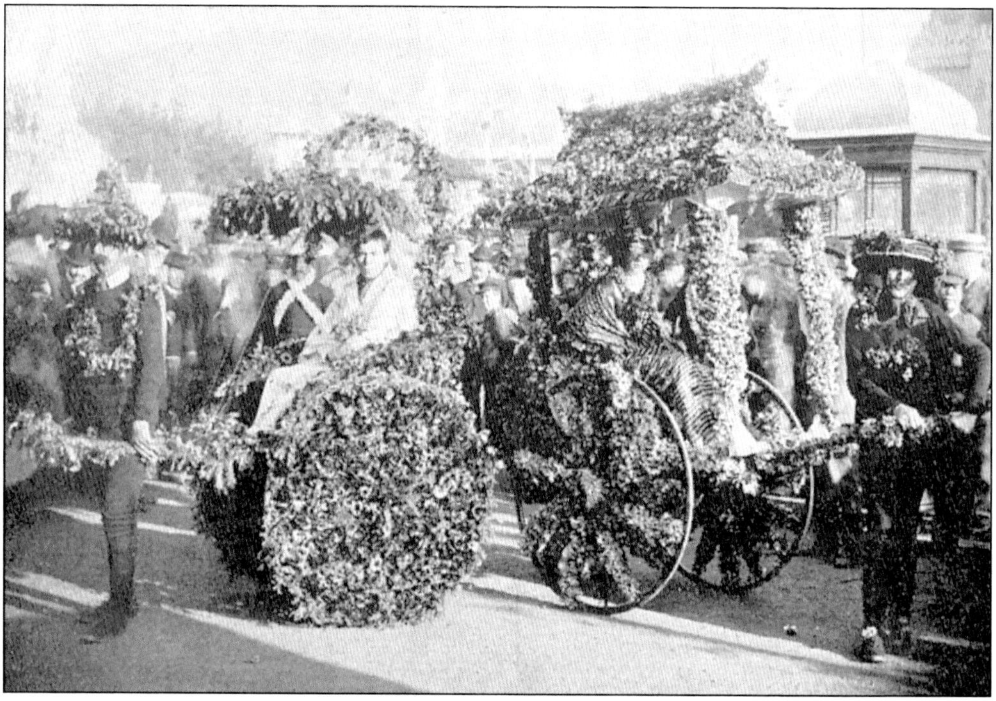

Japanese Day featured a parade of beautifully decorated jinrikishas, each of a different design. Some of the vehicles, passengers, and attendants were so festooned with flowers that they barely could be seen. The event was a feast for every spectator's sense of smell, sight, and wonder. Other named days included Independent Order of Good Templars' Day (February 12); Mystic

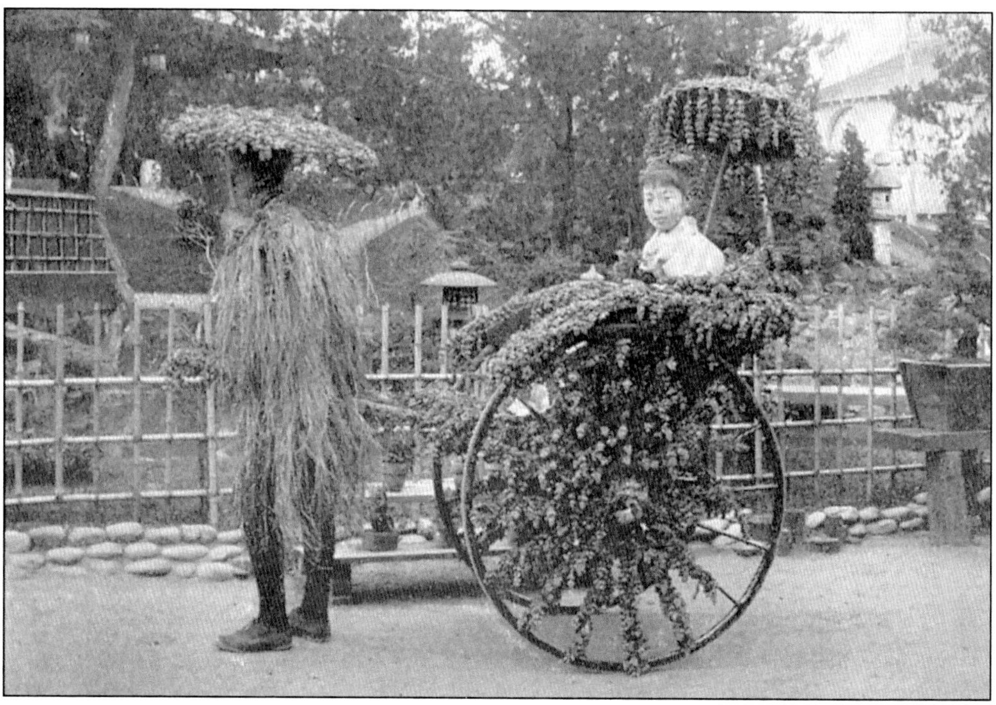

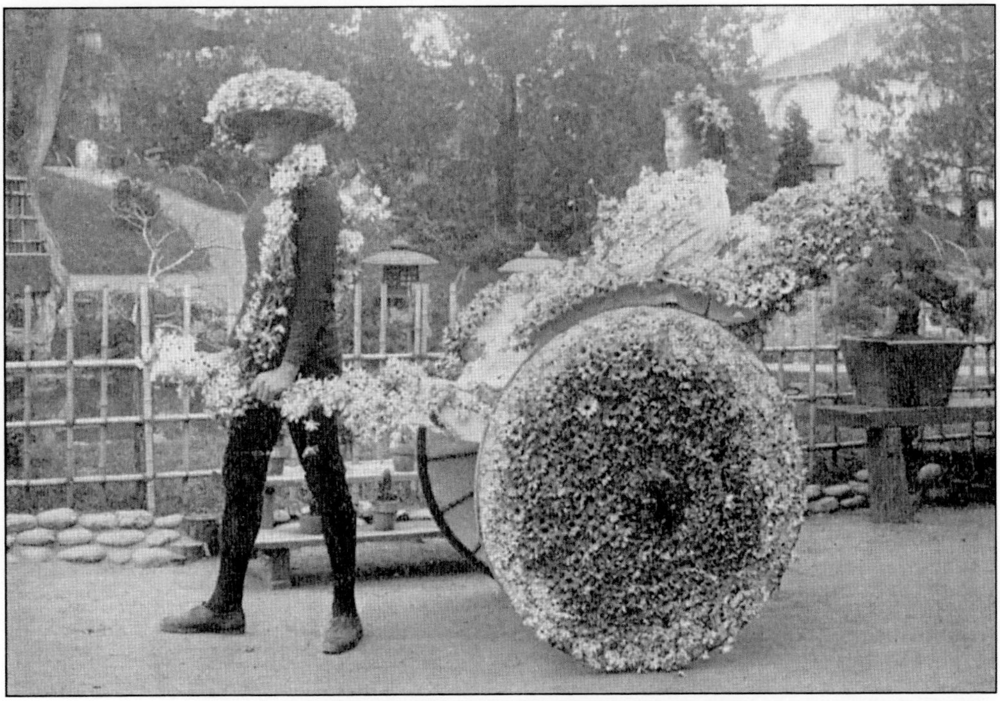

Argonauts of the Golden Gate Day (March 19); Santa Cruz Day (March 21); Canadian Day (April 2); Order of Chosen Friends Day (April 12); Poet's Day (April 23); Benevolent Order of B'nai B'rith Day (May 10); Women's Christian Temperance Union Day (May 11); Viticultural Day (May 11); Miners' Day (May 22); and Afro-American Day (June 5).

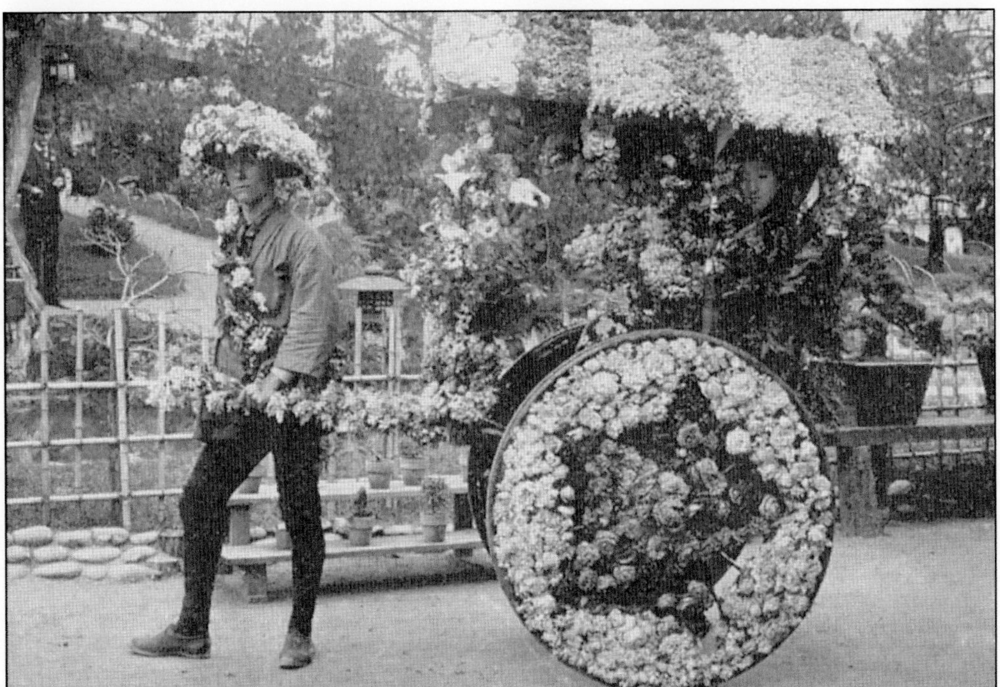

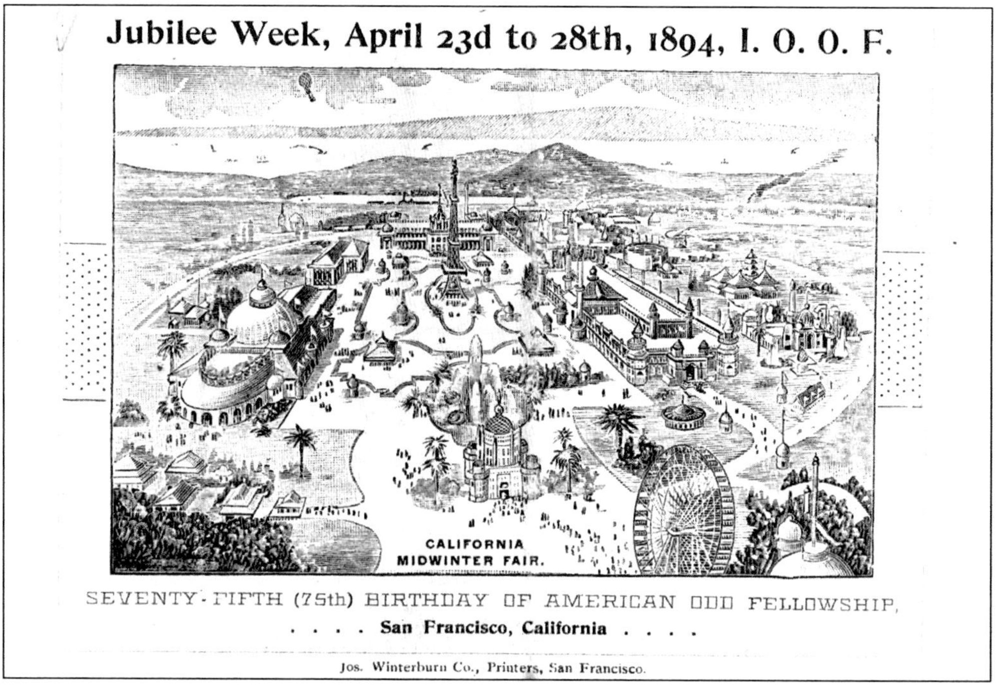

The Midwinter Fair Congresses considered "timely questions in the line of moral, social, and industrial advancement." Separate conferences considered economics and politics, religion, education, geography, women's issues, temperance, and literature, among others. Most groups met away from the fairgrounds in Golden Gate Hall in downtown San Francisco. The International Order of Odd Fellows used their own hall, then as now at Seventh and Market Streets, for their April encampment.

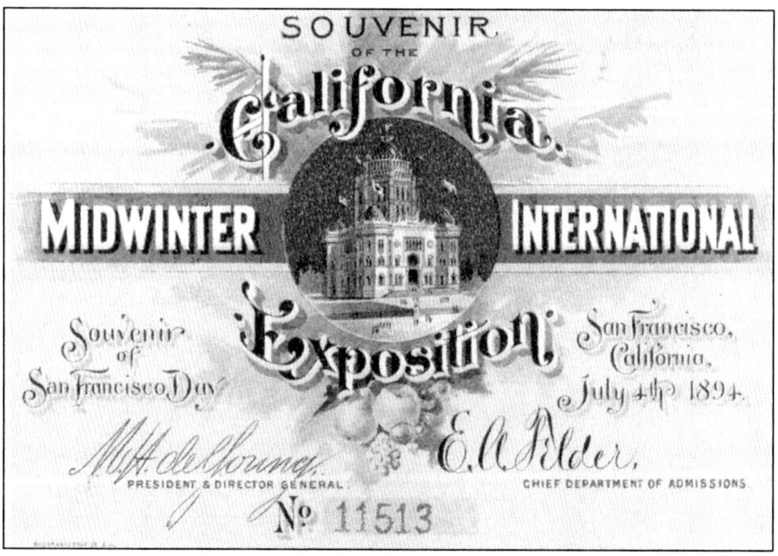

San Francisco Day was July 4. The souvenir admission tickets for the occasion pictured the Administration Building on their face and the main exhibition halls on their reverse. To avoid any misunderstanding about their keepsake status, the ticket proclaimed it twice. Apparently fairgoers got the message and saved them. All the Exposition's souvenir tickets are very rare, but this is the most common.

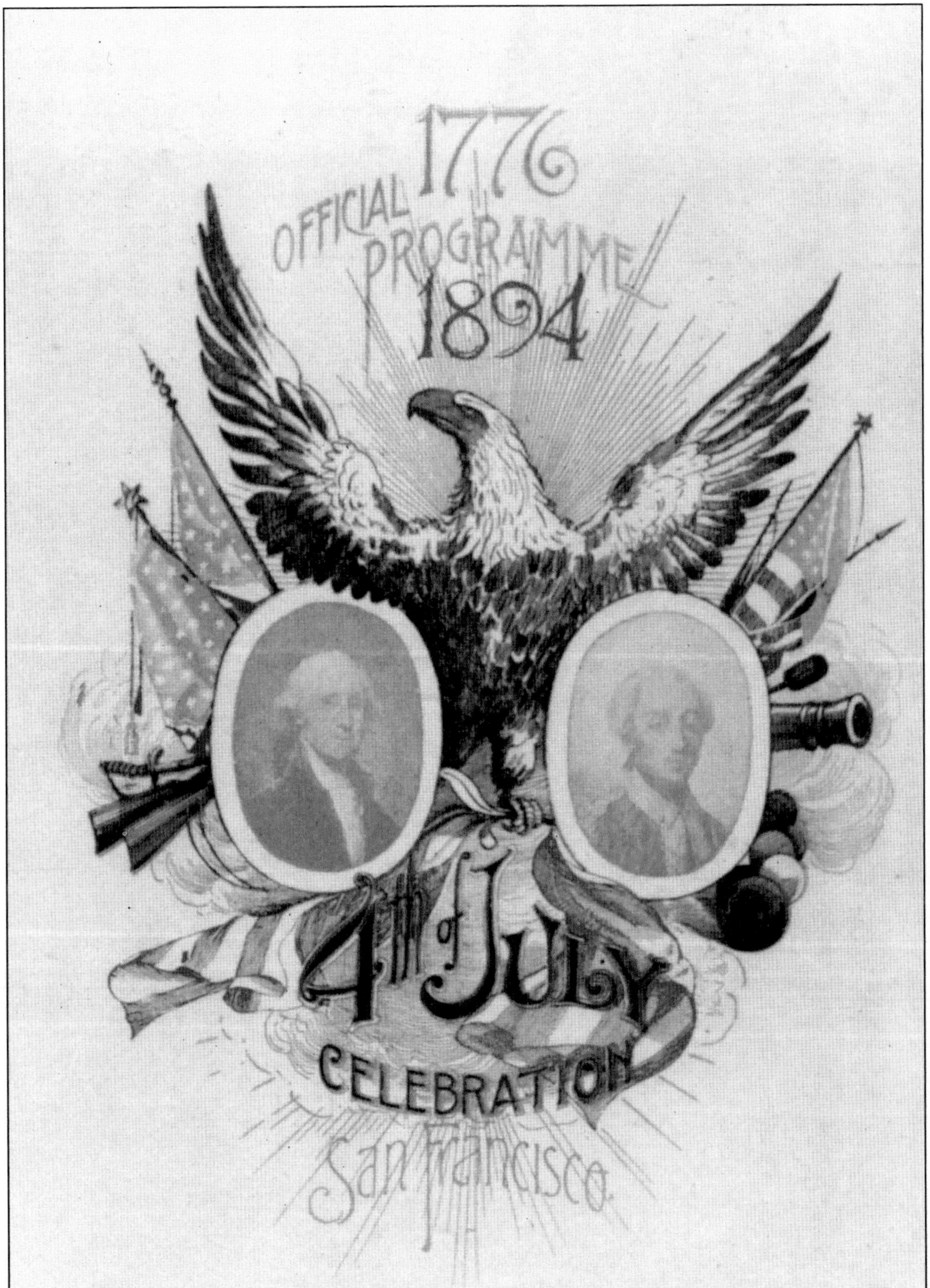

July 4th combined three celebrations into one grand observance. Not only was it San Francisco Day, it also celebrated Independence Day and Sunset Day, which marked the Fair's finale. Naturally, the festivities began with a parade to the fairgrounds. Literary exercises followed in Festival Hall. Fireworks—31 separate displays—ended the day and signaled the official end of the Fair. More than 79,000 attended, the Exposition's second largest attendance.

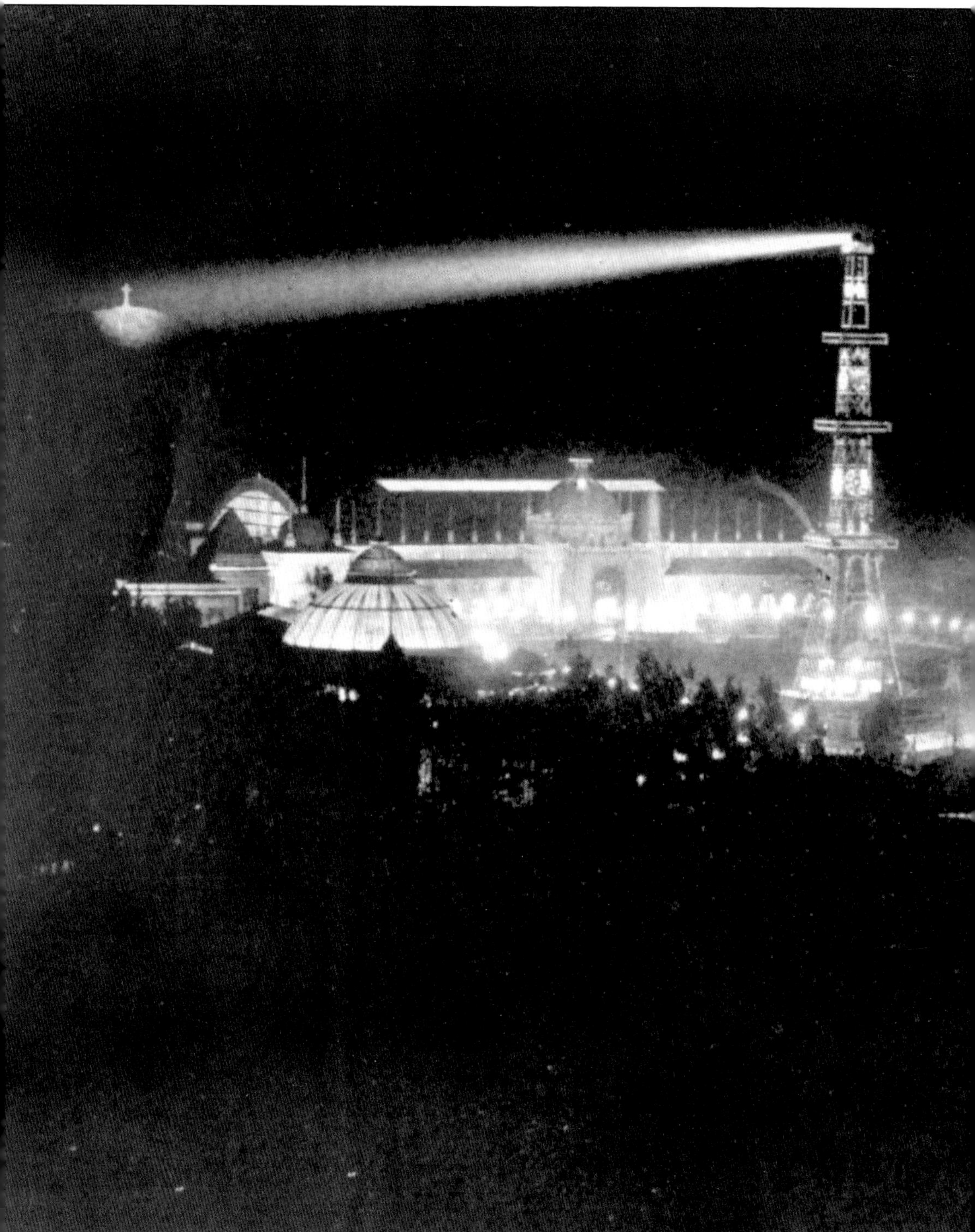

For the next few nights, while exhibitors and vendors packed their goods and vacated the premises, the grounds remained open, free to the public. Some attractions still operated, but most no longer did. On the night of July 8, the Mexican Band, no longer under contract but still

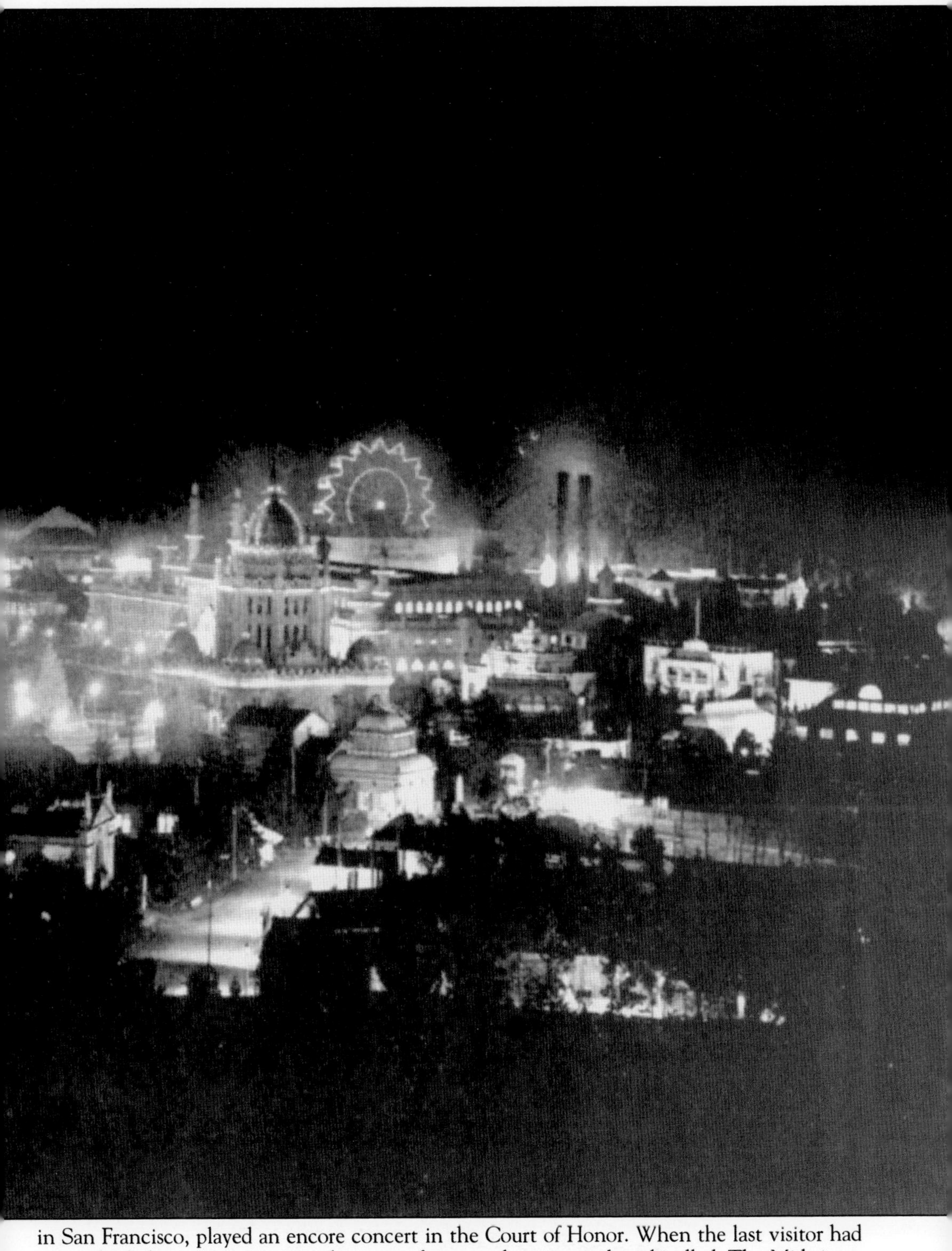

in San Francisco, played an encore concert in the Court of Honor. When the last visitor had gone, the lights went out across the entire fairgrounds, never to be rekindled. The Midwinter Fair, a fantasy only a year earlier, was now a memory.

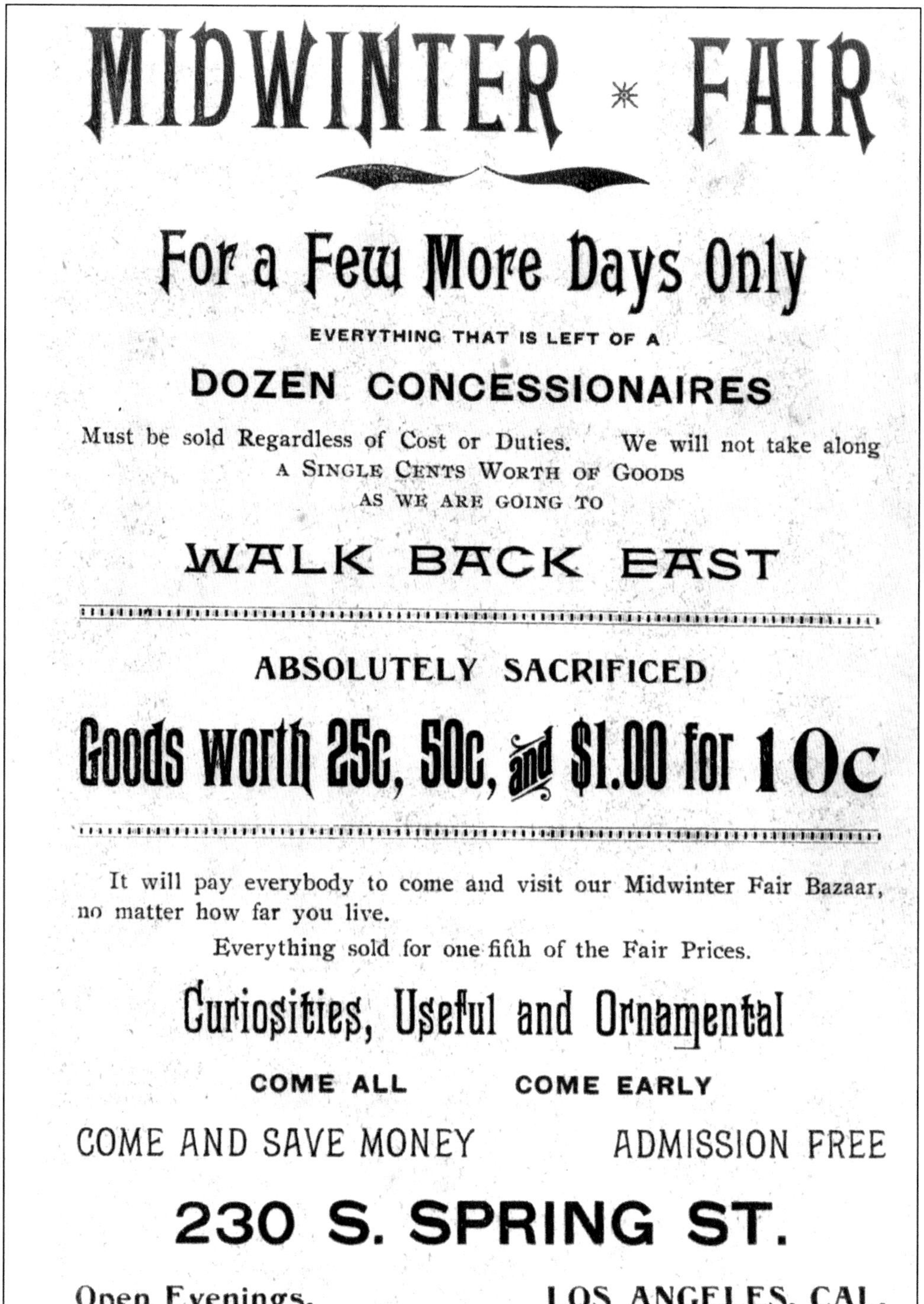

Many merchants liquidated their unsold goods. The handbill here was printed on the blank back of a page recycled from a souvenir album. The books probably had so little market value at the time that it cost less to cut them up than to buy new paper. Curiously this sale took place in Los Angeles, not San Francisco.

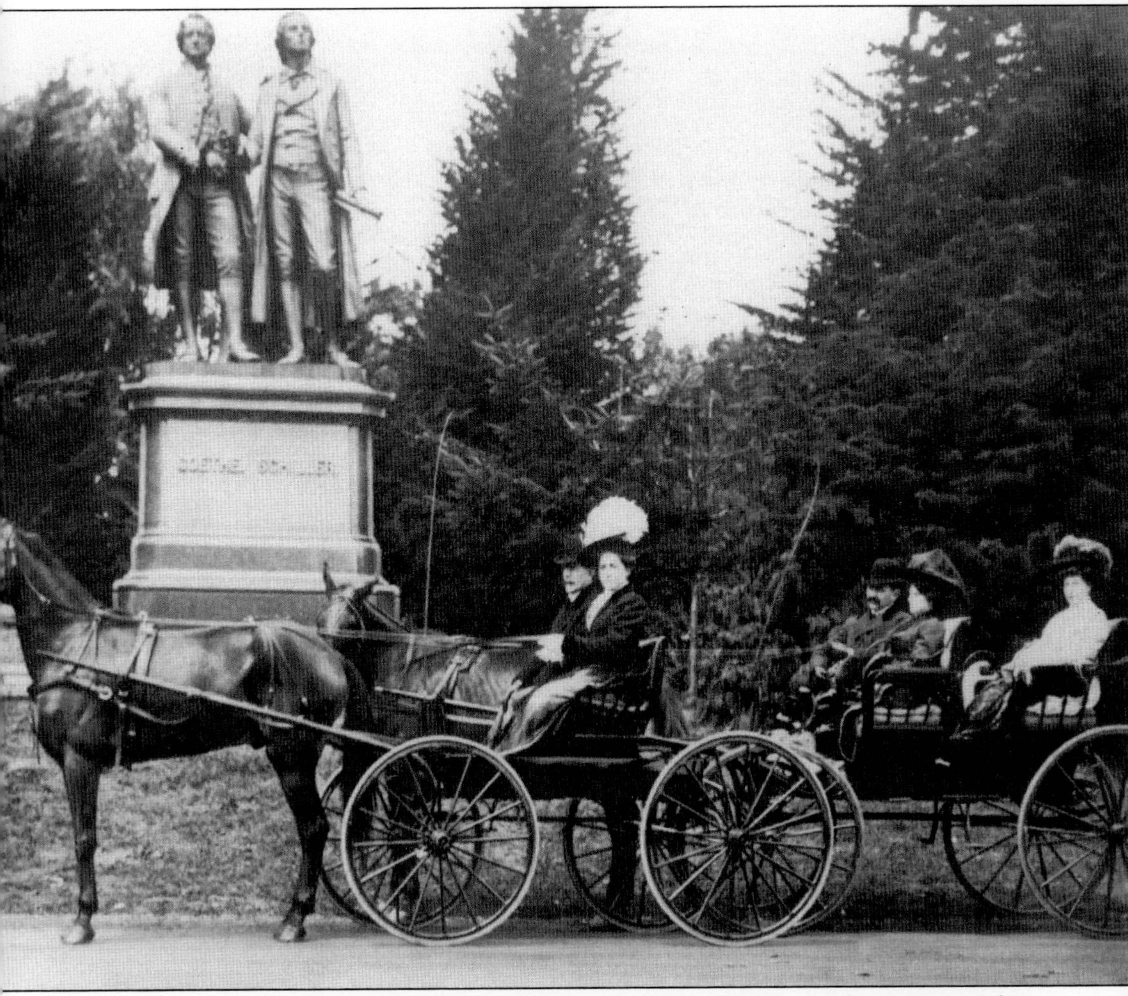

After the Fair, Park Superintendent John McLaren took his revenge on the usurper in his park. The Fair's management had agreed to restore the grounds, but they didn't. McLaren and his crew tore down the buildings themselves and sold the lumber. They dug up the concrete foundations, then hauled them to the dump. When no one responded to his requests to remove the Electric Tower, which he feared would become a permanent observation platform, McLaren dynamited it into junk, sold the scrap, and put the proceeds into the Park Improvement Fund. The Fine Arts Building reopened on March 23, 1895, San Francisco's first municipal art gallery. De Young presided at the transfer of title ceremonies. He wrote later, "I formally presented the museum to the park with the understanding that it was to remain in Golden Gate Park under the title of the *Memorial Museum* and to be open free every day in the week. Its doors were then thrown open." The same year the Goethe-Schilling monument, seen here, was installed on the lawn where the Horticulture Building had been.

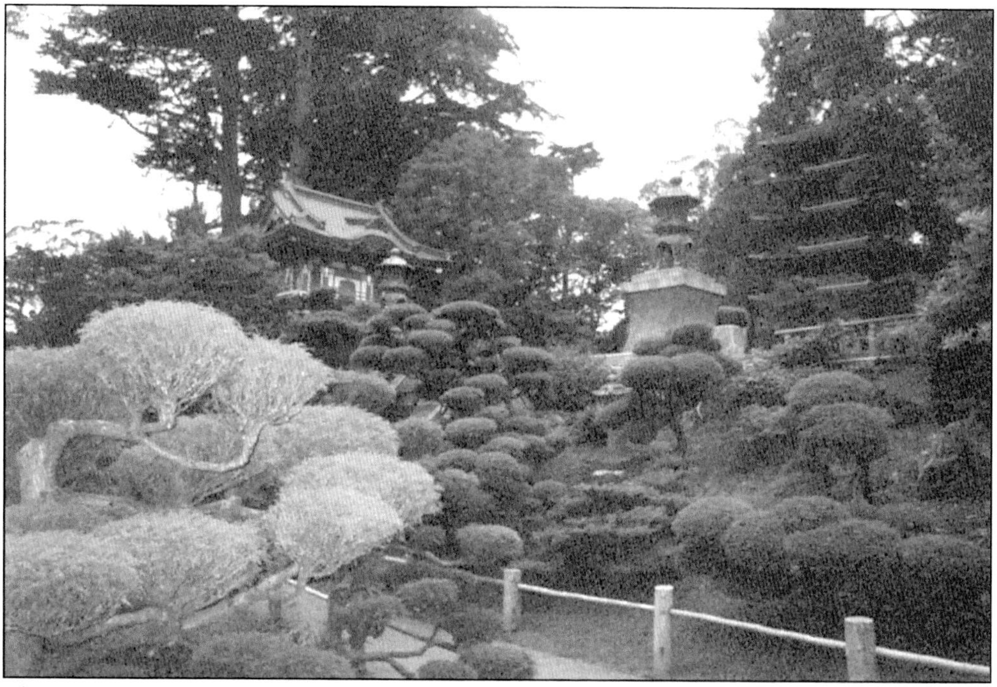

The only structures McLaren spared were the Fine Arts Building—always intended as a permanent addition; the Royal Bavarian Pavilion, which for a time was an annex of the new museum; and the Japanese Village, now the Japanese Tea Garden seen here. The steps and terraces of the Court of Honor, walkways, and underpasses also remained.

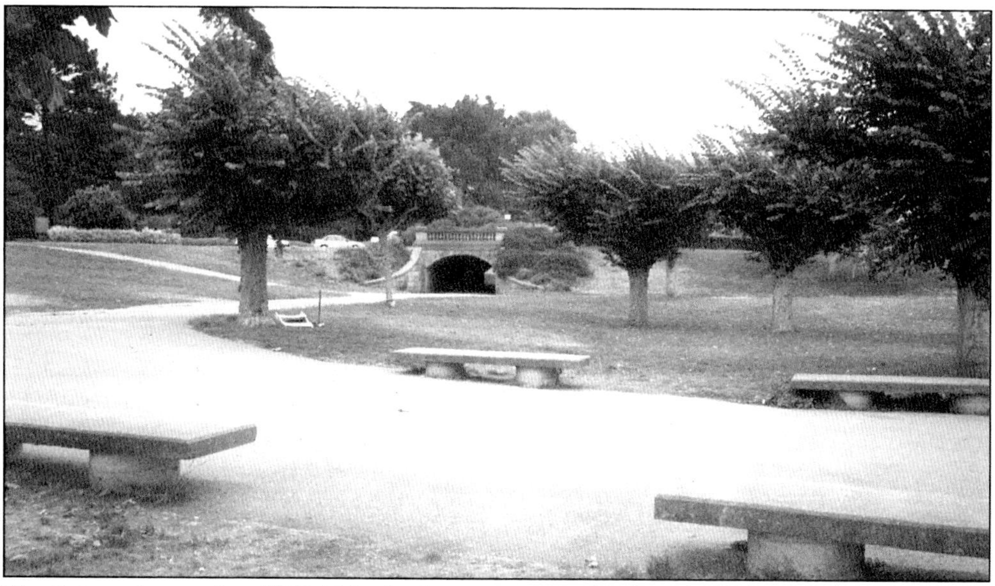

The Fair made a profit of $32,464, which everyone involved credited the rigorous system that affirmed vendor sales; clerks conducted more than 76,000 audits. The funds were used primarily to buy items for the new municipal art gallery. Together with the cost of the Fine Arts Building and the Court of Honor, statuary, benches, and miscellaneous additions such as the underpass seen here, the Exposition donated almost $200,000 to the park.

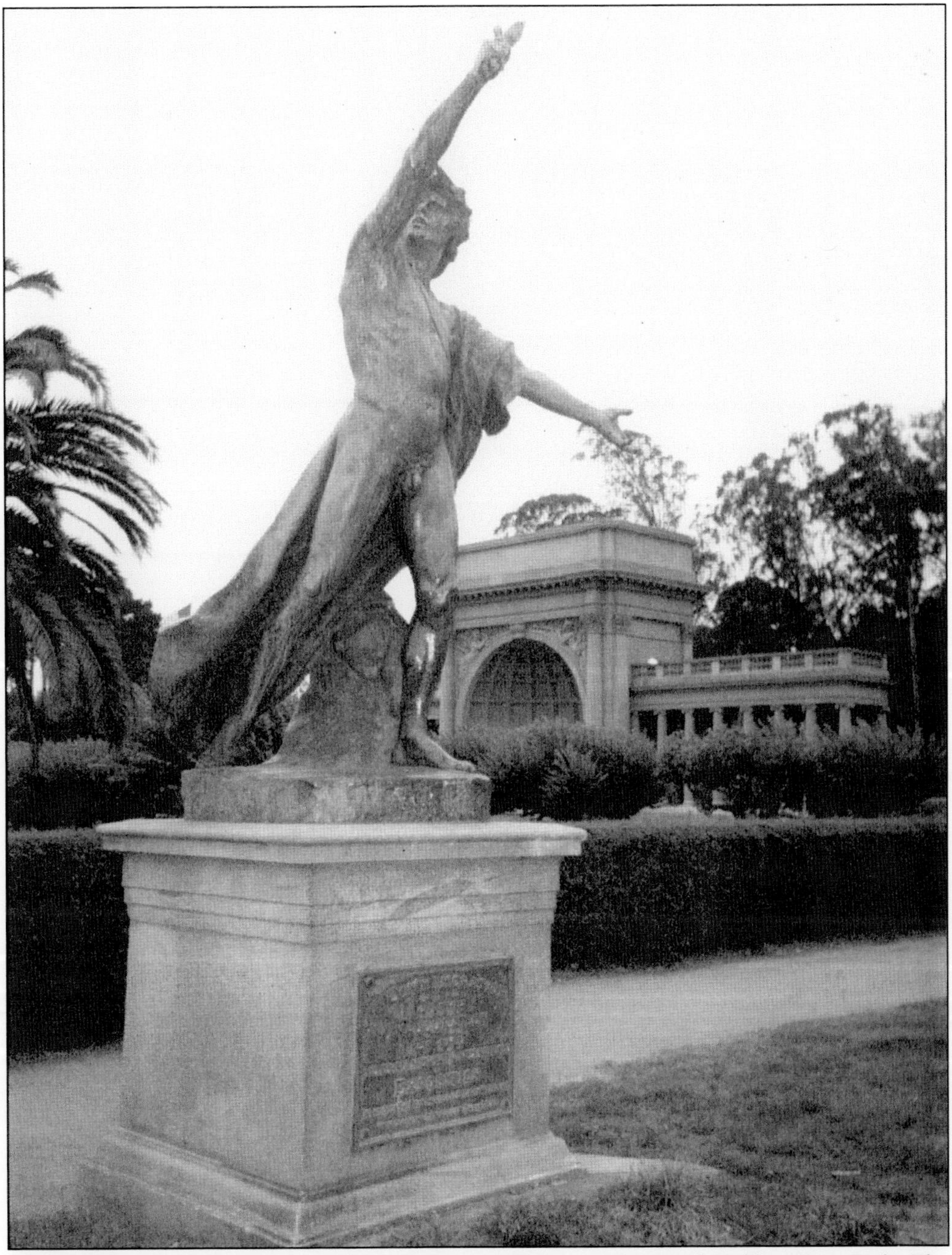

Today there are still reminders of the Midwinter Fair in and around Golden Gate Park. The neighborhood south of the fairgrounds took its name from the Sunset City, the only one ever named for an international exposition. The Japanese Tea Garden, the Cider Press, the general design of the Court of Honor, and some miscellaneous construction endure. A statue exhibited at the Fair of King Leonidas of Sparta, holding aloft a flaming sword, marks the approximate site of the ground breaking; a plaque at the base provides this information to passersby. Another plaque in the Tea Garden explains that the original entry doors, since replaced, were from the Fair. The Fine Art Building's sphinxes still sit inscrutably where they have been for many years.

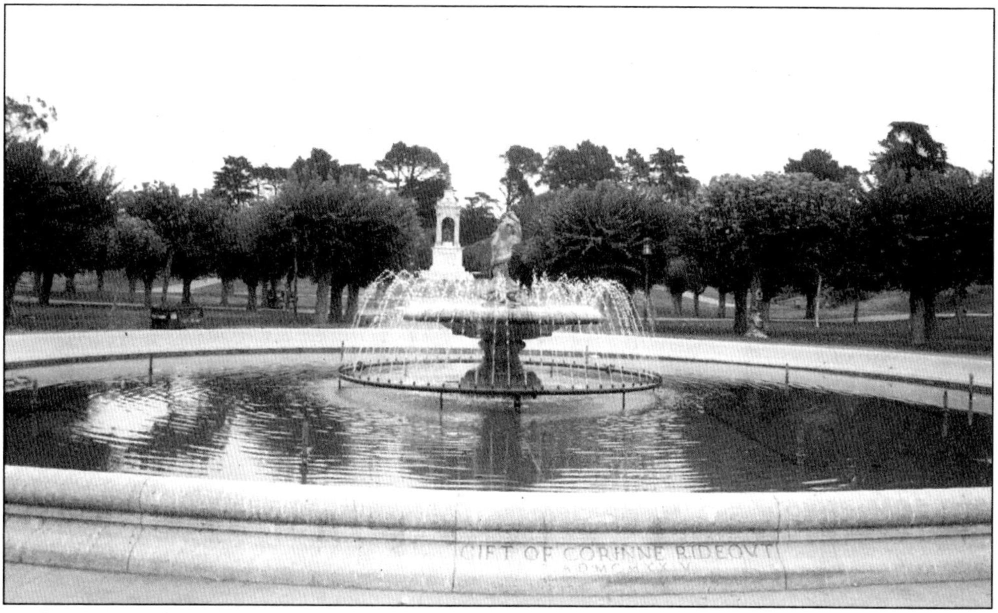

In the Court of Honor—now the Concourse—the band shell stands approximately where the Administration Building was. A fountain, foreground, sits where the Electric Tower once stood. A newer fountain has replaced the Electric Fountain. In the background, the Francis Scott Key monument stands were the Manufacturers Building was once.

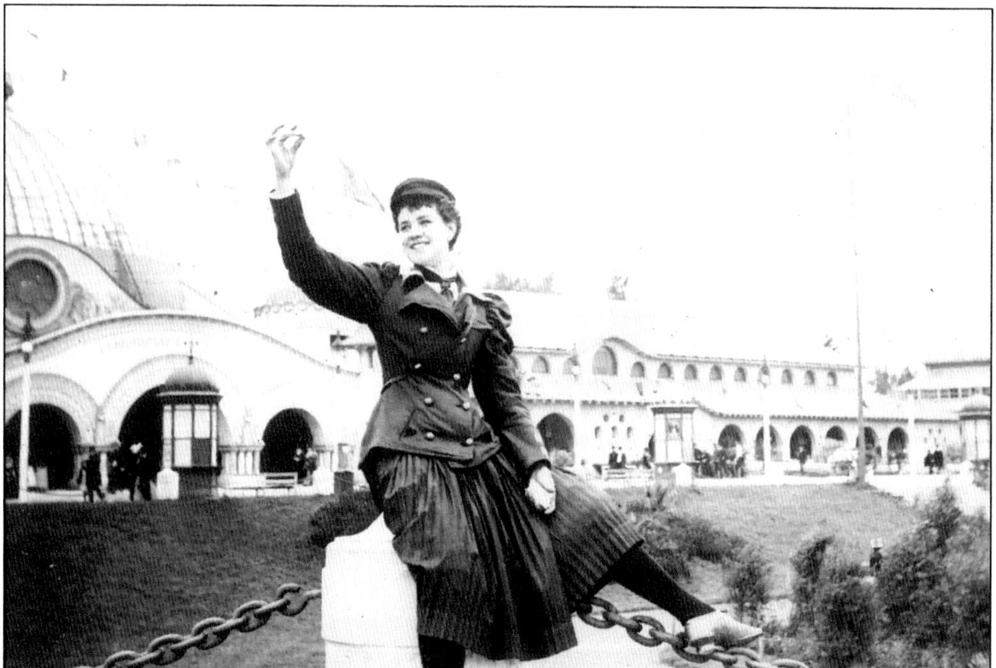

The Fine Arts Museum was torn down in 1928. Named for Director-General de Young, its successor also has been razed, to be replaced with a complex that looks like nothing so much as three prisoner of war barracks and a guard tower. It seems there will be a tall building in Golden Gate Park after all and for many years to come.